"The great, escalating crises of climate disruption and biodiversity loss are unavoidably entangled with the population problem. Trevor Hedberg analyzes that problem rationally and ethically and proposes practical solutions. His efforts yield conclusions that are logical yet humane and realistic yet hopeful."

John Nolt, Distinguished Service Professor Emeritus of Philosophy and Research Fellow in the Energy and Environment Program at the Howard H. Baker Jr. Center for Public Policy, University of Tennessee, US

"Philosophers and ethicists have begun to enter the discussion of overpopulation in recent years, but it has been slow and piecemeal. Until now. Trevor Hedberg's book is an impressively thorough investigation into the ethical and policy issues raised by our ever-increasing numbers."

Travis N. Rieder, Director of the Master of Bioethics degree program, Johns Hopkins Berman Institute of Bioethics, US

"It's taboo to discuss reproductive issues in the West except in the context of liberty. As a result, Western thought has failed to reflect on the deepest problem of our age – the collision with planetary boundaries. Hedberg's *Overpopulation* is a breath of fresh air. It connects the dots, asks the right questions, and leads to new insights."

Martin Schönfeld, Professor of Philosophy and Faculty in the College of Global Sustainability, University of South Florida, US

"Trevor Hedberg summarizes the most credible empirical evidence and best moral arguments regarding population growth. He also presents original arguments that are subtle and convincing. Furthermore, the book is accessibly written, clear, and persuasive. It's definitely a must read for anyone interested in this topic!"

Ramona C. Ilea, Professor and Philosophy Department Chair, Pacific University, US

The Environmental Impact of Overpopulation

This book examines the link between population growth and environmental impact and explores the implications of this connection for the ethics of procreation.

In light of climate change, species extinctions, and other looming environmental crises, Trevor Hedberg argues that we have a collective moral duty to halt population growth to prevent environmental harms from escalating. This book assesses a variety of policies that could help us meet this moral duty, confronts the conflict between protecting the welfare of future people and upholding procreative freedom, evaluates the ethical dimensions of individual procreative decisions, and sketches the implications of population growth for issues like abortion and immigration. It is not a book of tidy solutions: Hedberg highlights some scenarios where nothing we can do will enable us to avoid treating some people unjustly. In such scenarios, the overall objective is to determine which of our available options will minimize the injustice that occurs.

This book will be of great interest to those studying environmental ethics, environmental policy, climate change, sustainability, and population policy.

Trevor Hedberg is a postdoctoral scholar at The Ohio State University, USA, jointly affiliated with the College of Pharmacy and the Center for Ethics and Human Values.

Routledge Explorations in Environmental Studies

The Role of Non-state Actors in the Green Transition
Building a Sustainable Future
Edited by Jens Hoff, Quentin Gausset and Simon Lex

The Creative Arts in Governance of Urban Renewal and Development
Rory Shand

What Can *I* do to Heal the World?
Haydn Washington

Towards a Society of Degrowth
Onofrio Romano

Romantic Anti-capitalism and Nature
The Enchanted Garden
Robert Sayre and Michael Löwy

Liberty and the Ecological Crisis
Freedom on a Finite Planet
Edited by Christopher J. Orr, Kaitlin Kish, and Bruce Jennings

Environmental Justice and Oil Pollution Laws
Comparing Enforcement in the United States and Nigeria
Eloamaka Carol Okonkwo

The Environmental Impact of Overpopulation
The Ethics of Procreation
Trevor Hedberg

www.routledge.com/Routledge-Explorations-in-Environmental-Studies/book-series/REES

The Environmental Impact of Overpopulation
The Ethics of Procreation

Trevor Hedberg

LONDON AND NEW YORK

First published 2020
by Routledge
2 Park Square, Milton Park, Abingdon, Oxon OX14 4RN

and by Routledge
605 Third Avenue, New York, NY 10017

First issued in paperback 2021

Routledge is an imprint of the Taylor & Francis Group, an informa business

© 2020 Trevor Hedberg

The right of Trevor Hedberg to be identified as author of this work has been asserted by him in accordance with sections 77 and 78 of the Copyright, Designs and Patents Act 1988.

All rights reserved. No part of this book may be reprinted or reproduced or utilized in any form or by any electronic, mechanical, or other means, now known or hereafter invented, including photocopying and recording, or in any information storage or retrieval system, without permission in writing from the publishers.

Trademark notice: Product or corporate names may be trademarks or registered trademarks, and are used only for identification and explanation without intent to infringe.

Publisher's Note
The publisher has gone to great lengths to ensure the quality of this reprint but points out that some imperfections in the original copies may be apparent.

British Library Cataloguing-in-Publication Data
A catalogue record for this book is available from the British Library

Library of Congress Cataloging-in-Publication Data
A catalog record has been requested for this book

ISBN 13: 978-1-03-223676-6 (pbk)
ISBN 13: 978-1-138-48975-2 (hbk)

DOI: 10.4324/9781351037020

Typeset in Bembo
by Wearset Ltd, Boldon, Tyne and Wear

For Donald Hatcher

Contents

Preface xi

PART I
Confronting the problem 1

1 The uncomfortable reality of rising numbers 3

2 The gravity of the population problem 14

PART II
Intergenerational ethics, population policy, and personal procreative obligations 31

3 Intergenerational equity and long-term environmental impacts 33

4 The moral duty to halt population growth 49

5 Policies that promote smaller families 63

6 Individual procreative obligations 84

PART III
Objections from alternative approaches to procreative ethics 107

7 Antinatalism 109

8 Reproductive rights and procreative freedom 131

x *Contents*

PART IV
Lingering questions   141

9 What about immigration?   143

10 What about the nonhuman community?   151

11 Can we solve the problem?   161

    Appendix: the non-identity problem   167

    Index   179

Preface

When I was born, there were five billion people in the world. In the time since, we have added more than 2.5 billion people to that tally. As a graduate student studying applied moral philosophy, I was baffled that people were not saying more about this subject and its connection to climate change, biodiversity loss, and a variety of other environmental problems we now face. How could philosophers – people who specialize in challenging conventional wisdom – be so silent on social norms regarding procreation when there seemed to be so much evidence that our burgeoning numbers were contributing to such significant problems?

That silence appears to have ended. During the last five years, candid discussions of overpopulation and the ethics of procreation have resurfaced in many disciplines. While there remains disagreement about the severity of the problem and what should be done about it, I view the renewal of this discussion as a step in the right direction. We cannot address such complex issues by ignoring them or subverting the conversation out of a fear that we might say something politically incorrect. I am certain this book will not be the last one addressing population growth, but I hope that it will, like the works that have preceded it, improve our understanding and push the conversation in new and promising directions.

One of the recurring themes of this book is that moral problems of a grand scale usually do not allow for tidy solutions. However we try to address these problems, there will almost always be winners and losers. Some will benefit; others will be harmed. The key is determining which outcome is most justifiable when none of the options available lead to an ideal resolution. I believe our current circumstances with respect to population growth fit this description. If we do not do anything to decelerate population growth and (in the long term) pursue population reduction, there will be devastating consequences for future people. However, if we do pursue certain measures to slow and eventually reverse global population growth, presently existing generations may suffer harm or injustice in the near term. So how do we balance these competing moral considerations? At its core, this book is an attempt to answer that question.

The good news is that there are measures that we can take to address population growth that fall far short of draconian one-child policies or forced sterilizations. There are also ways to mitigate the undesirable impacts associated with dropping fertility rates. Some of these strategies will not be costless, but they are costs that I think we should bear. I do not hold this view because I think the world is on the brink of destruction or that the ecological state of the planet is becoming unsalvageable. This is not a manifesto about how humanity has destroyed the natural world or stands poised to doom future generations to lives of misery. My view follows simply from a combination of empirical observations and a few moral principles that I hold to be true. This book strives to be a call to action without being a doom-and-gloom narrative, though readers unfamiliar with the facts may still find them a little unsettling.

In working through these issues over the last several years, I owe a great debt to those who have influenced my thinking on this subject and on ethics more generally. Since many of the ideas developed here were first explored in my doctoral dissertation, I must first extend thanks to the members of my dissertation committee who helped me develop my initial thoughts into a coherent monograph: John Nolt, David Reidy, Jon Garthoff, and Phil Cafaro. John Nolt deserves special thanks for his painstaking comments on early drafts of that material and for ongoing discussions about these issues throughout my graduate career. I also extend thanks to Marcus Arvan, Bob Fischer, Stephen Gardiner, Ben Hale, Donald Hubin, Carter Hardy, Laura Kane, Corey Katz, Alex Levine, Martin Schönfeld, Piers Turner, and Eric Winsberg for feedback on work in progress related to this project.

I also had the opportunity to present freestanding papers and chapter drafts at a variety of professional venues while working on this manuscript. I thank audiences at the University of South Florida, University of Tampa, University of Tennessee, University of Washington, and the 2017 meeting of the Association for Practical and Professional Ethics for comments and criticisms related to this material.

I have also had significant help with the preparation of this particular manuscript. I want to thank three anonymous reviewers for Routledge for their detailed comments on the original book proposal. That feedback helped me improve the structure of the book in some important ways. I must also thank Annabelle Harris and Matt Shobbrook for helping me prepare and edit the manuscript and for their patience during the writing process.

Finally, I must extend a special thanks to clinical psychologist Wendi Born, who helped me cope with a major depressive episode that occurred midway through working on this manuscript. If I had not recovered from that spell, I doubt this book would have been finished. I am also grateful to the friends and family who offered their support during that time, however small it may have seemed.

Part I
Confronting the problem

1 The uncomfortable reality of rising numbers

As I write this sentence, our planet's human population sits at about 7.8 billion people. A century ago we were only at 1.8 billion, and it took us all of human history until 1804 to reach 1 billion.[1] This unprecedented surge in the human population over the last century has resulted from a combination of significant technological advancements. Edward Jenner's smallpox vaccine, which was developed in 1796, helped to prevent 2–3 million annual deaths. Further developments in medicine increased life expectancy and reduced infant mortality rates. Advancements in agriculture also enabled us to feed larger groups of people. While these changes generally made human lives better, they also created conditions where our numbers could increase more rapidly than ever before.

Now, two decades into the twenty-first century, we face a series of moral challenges intertwined with our population size. The average global temperature is rising, and climate change threatens the welfare of present and future people alike. Species are rapidly going extinct, raising the specter of irreplaceable losses of beauty, biological knowledge, and environmental resources. Regional food and water shortages loom across the world. Population size in isolation does not necessarily cause any of these problems, but it contributes to all of them. We cannot address these issues adequately without considering the role that population size plays in their continuation.

Yet, while the importance of human numbers in these matters strikes some as obvious, not everyone shares that perception. Even those who care deeply about climate change, nonhuman animals, and the environment do not always view population growth as a central cause of our environmental problems or as something that we ought to address. Some are skeptical that global population will continue on its current upward trajectory for much longer (Pearce 2010; Bricker and Ibbitson 2019). Others believe that economic patterns entail that scarcities created by population growth are ultimately resolved (Simon 1996). Even among the environmentally conscientious, discussions of population are sometimes avoided because of their rhetorical challenges and worries about the fraught, morally unsettling history of previous population policies (Roberts 2018).

In the pages that follow, I will argue that these views are mistaken: population is a serious contributor to our environmental problems, we are morally obligated to pursue the swift deceleration of population growth, and there are morally permissible means of achieving this outcome – means that avoid the coercive measures employed in the past. The path to defending these conclusions starts with understanding our current circumstances. I begin by addressing the question of why population discussions have been so rare in the last three decades despite the topic's importance. Once this background context is established, I highlight some important aspects of the book's framework and outline how the remaining chapters will proceed.

Breaking the silence on population

While there has been an upswing in academic discussions of population growth recently (e.g., Clarke and Haraway 2018; Conly 2016; Coole 2018; Crist 2019; Rieder 2016), this trend was preceded by two decades of near silence on the subject, and even now, direct discussion of the topic in the political domain remains extremely rare. Mainstream media often ignores or understates the significance of population growth, and many environmental organizations and institutes – the Jane Goodall Institute, The Nature Conservancy, and the Rainforest Action Network among them – do not acknowledge the contribution of overpopulation to the environmental problems that they are trying to resolve (Shragg 2015, pp. 23–32). If population growth really is a noteworthy problem, then why did people stop talking about it in the 1990s even as global population continued to rise?

The retreat from the population question originated from a combination of factors (Campbell 2012). First, with few exceptions, fertility rates around the world have been declining since the 1960s. Even now, many nations are still above the threshold of replacement level fertility, which is about 2.1 births for woman, but a few decades ago, dropping fertility rates created the impression that the population problem was resolving itself. Moreover, as fertility was dropping, patterns of overconsumption started to become more visible, especially as concern about climate change grew. Developed nations have generally consumed far more energy and resources than developing nations, and as a result, they have been (and continue to be) the primary emitters of greenhouse gases.[2] The high-consumption lifestyles of those in wealthier nations have other powerful effects on the environment as well, such as pollution, deforestation, and biodiversity loss. These impacts are easier to see than the subtler effects of gradual population rise. Thus, the focus on reducing consumption has tended to eclipse concern about stabilizing and reducing population.

Declining fertility and a focus on reducing consumption were not the biggest factors, however. The most significant development in removing population growth from policy discourse occurred at the 1994 United Nations International Conference on Population and Development (ICPD) held in Cairo, Egypt. ICPD, in contrast to previous population conferences held by the UN, emphasized the needs of women around the world. Prior to the conference, discussion of rising population and the connection between population growth and environmental destruction became politically incorrect because suggestions to stabilize or reduce population were perceived as disadvantageous to women. Recent history seemed to validate this perception: measures such as China's one-child policy and coercive episodes of family planning in India appeared morally unacceptable. Conference attendees wanted to distance themselves from these policies, and this desire led to a new strategy for addressing population issues: family planning and all other health-related issues related to women were combined under the label "reproductive health." While it was probably unintentional, the impression created by this change in language was that all family planning efforts prior to 1994 had been objectionably coercive. Those measures became associated with the derogatory label "population control" even though many family planning organizations established prior to ICPD made no efforts to limit or otherwise control fertility. Many of them aimed only at making family planning easier. Nevertheless, this false generalization has proven difficult to shake even decades later.

Martha Campbell (2012) highlights three other reasons that population discourse faded in the 1990s. For one, while these other developments were taking place, conservative think tanks and religious leaders opposed to abortion and family planning managed to reduce the attention being paid to population growth. The general strategy of these groups was to reinforce the notion that world population growth is at an end (Lutz, Sanderson, and Scherbov 2001). This idea has gained a foothold in the media and diminished the public's concern for population growth.[3] Additionally, the AIDS epidemic in Africa drew a lot of attention, and many believed that it would reduce population growth in the region significantly.[4] Finally, classic demographic transition theory suggests that people have to be coaxed by changes in society to want a smaller family (Potts and Campbell 2005, pp. 180–181). According to demographic transition models, people are naturally inclined to have high birth rates until their societies develop from a pre-industrial to industrialized economic system. While this phenomenon has many exceptions, its general acceptance led people to believe that we cannot incentivize people in the developing world to have fewer children without unjust forms of persuasion or coercion.

Thus far, I have highlighted six different factors that contributed to avoiding discussion of population growth, but there is one more worth mentioning. Population growth tends to be associated with economic growth, and as I will discuss in later chapters, population shrinkage can lead to short-term economic challenges. Thus, at least in economics, population growth tends to be viewed in a positive light. Given that economic growth is a priority in many societies, political leaders are not always motivated to promote measures that could reduce birth rates.

6 *Confronting the problem*

We now have a fairly comprehensive account of why the population question has been neglected in recent decades. We should acknowledge that this neglect was in part motivated by sensible moral concerns. The unsettling history of coercive population policies cannot be ignored, and those mistakes must be avoided in the future. A further worry is that population policies will be racially inequitable in their application. Since the countries that have the highest birthrates are predominantly in Sub-Saharan Africa and other areas of the developing world, non-white populations would be the most affected by any policy placing restrictions on procreation. Some also worry that focusing on population's contribution to environmental problems will shift the focus away from the high-consumption lifestyles of the wealthy. Since reducing the consumption rates of those in the wealthiest parts of the worlds will be an essential part of responding to our environmental problems, it is reasonable to keep that objective as a top moral priority.

Unfortunately, limiting discussion of population may have done more harm than good. Following ICPD, access to family planning options did not expand sufficiently to accommodate the increasing numbers of women who wanted them, and the term "reproductive health" was more difficult for governing bodies and the general public to understand and support than the narrower term "family planning" (Campbell 2012, pp. 47–48).[5] Additionally, dismissing population growth created new problems for reducing overall consumption. The effects of population growth can be hard to detect, so people often fail to notice the ways that population growth undermines the efficacy of reducing consumption rates.[6]

To see how population growth can subtly undermine efforts to reduce total environmental impact, consider efforts to preserve rivers in the United States.[7] During the first half of the twentieth century, thousands of dams were built across the United States. Once people realized that these dams caused considerable damage to local ecosystems, many tried to preserve the best remaining rivers in the nation. Demand for water was still rising, but instead of taking water from others or creating more dams to make additional water available, people tried to make more efficient use of the water that was already available. From 1980 to 1995, per capita use of water in the United States decreased by 20 percent (Jehl 2002). Their efforts to reduce consumption were successful, but those efforts were undone by rising population. The United States population grew by 16 percent during the same 15-year time period, so the progress toward solving the problem was negligible: the need for water was virtually as great in 1995 as it was in 1980 despite the reduction in the water consumption per person.

The moral of this short history lesson is that improvements to efficiency in our use of resources are solutions only to the extent that they outpace population growth. Reduced rates of consumption are only temporary solutions in the context of an ever-increasing population: if numbers continue to rise, then eventually new solutions will be needed. To reiterate an earlier point, there is no doubt that we must reduce our consumption rates to avoid perilous climate change and a host of other catastrophes, but reducing each person's greenhouse gas emissions, energy consumption, water use, and other resource consumption will not amount to sustainable living if population growth continues unchecked. Some even claim that we must ultimately reduce global population to about two billion to maintain an adequate to comfortable standard of living in the long term (Smail 1997; Foreman 2012). I will return to that estimate in the next chapter. For now, what matters is recognizing that we should not ignore the problem any longer. We have to assess what should be done to promote population stabilization and reduction if we are to adequately respond to the moral challenges that await us in the remainder of the century.

Moral principles and reflective equilibrium

Throughout this book, I will be making claims about what people should do (or ought to do) in various circumstances. The "should" in this context denotes a moral duty or obligation. This is the same usage that applies when we tell children that they *should* respect their neighbors or that they *should not* tell lies. We all have some experience making moral judgments, so there is hopefully nothing mysterious about this concept. Where my approach differs from some other philosophers is that I will not be determining what we should do by relying solely on an established moral theory. The history of philosophy features several moral traditions that remain influential today, but any approach grounded in a single moral theory has two significant shortcomings: a limited audience and the problem of incompleteness.

Imagine for a moment that I told you I was going to adopt a utilitarian outlook on population growth. Utilitarianism is the moral theory that holds we should perform the action that, on balance, promotes the best overall outcome for everyone affected by the action, usually in terms of total happiness. If I adopt this approach, then anyone who disagrees with that view will be skeptical of my argument's success. Either I will have to concede that my argument is limited only to utilitarians, or I will have to give some compelling explanation as to why this moral theory is superior to all its competitors. I would rather not constrain my audience so severely, and since contemporary philosophers have reached no consensus about what moral theory is correct, I doubt that such an argumentative feat is achievable.[8] Even if I could demonstrate that one specific moral theory was correct, this book is not the place to attempt such a vast inquiry.

An additional problem with relying on a single moral theory is that none of them appear to capture the entirety of what is morally important. Almost anyone would agree that the consequences of our actions matter to whether that action is right or wrong, but far fewer are willing to say, as consequentialists do, that the probable outcomes of our actions are the *only* thing that determines whether our actions are right or wrong. Similarly, most will acknowledge that the development of admirable character is a central component of being a good person. Virtue ethics attempts to reduce morality to that task – the cultivation and expression of virtuous character traits (e.g., honesty, courage, empathy). But this does not seem to cover all important aspects of moral reasoning, especially since many of our modern moral problems are created by complex interactions of large groups of people. Some of these problems arise without anyone involved necessarily acting on malicious or otherwise unethical dispositions, so focusing on individuals' moral character may not be useful in determining what we should do.

I could catalog the incompleteness of other moral theories, but these observations are enough to motivate the general concern: at present no single moral theory appears to coherently capture everything that matters to our moral reasoning. I suspect one reason that several moral traditions have survived centuries of scrutiny that they each capture some central *component* of morality that competing theories fail to capture. If moral thinking is a jigsaw puzzle, they are all important pieces, but the task of completing the puzzle to see the finished picture still appears beyond our reach. Perhaps it is possible to synthesize all the plausible moral theories into a single unified framework, but thus far, no attempt at doing so has achieved anything close to a consensus among philosophers.[9] Moreover, I have no intention of crafting a unified theory of ethics in this text.

Given my concerns about relying on a single moral theory, I will instead approach the moral dimensions of human population growth using freestanding moral principles. These principles might be more commonly associated with some moral traditions than others, but the aim is to use principles that could be endorsed by any rational person engaged in serious moral reasoning, regardless of whether they have any leanings toward a particular moral theory. That ideal may prove too optimistic, though, so I will always provide a defense of any moral principle I present.

My method of justifying these moral principles involves trying to explain our considered judgments – those moral convictions that survive sustained critical reflection under conditions conducive to good reasoning (e.g., no manipulation, an absence of social biases) – in terms of moral principles that could be coherently unified. In philosophy, this strategy is commonly called *reflective equilibrium*.[10] With respect to moral reasoning, this method aims at a stable coherence between our considered judgments and the theoretical principles that explain those judgments. Full coherence often proves elusive, and so we often have to revise our moral principles or discard some of our old judgments. Hopefully, each time we do so, we get a little closer to a set of consistent and plausible principles.

One concern about reflective equilibrium is that it appears to favor our moral intuitions about various real-world cases, and it is not clear that our intuitions about these matters reliably lead to true belief. We can readily observe that moral intuitions vary across people and cultures, and some recent research suggests that our intuitions are grounded in automatic, unreflective responses (e.g., Haidt 2001; Greene et al. 2001). The variance among intuitions and the reflexive way in which they are often generated indicate that we should not depend heavily on them to conduct moral reasoning if our goal is the attainment of moral truths. At the same time, it is probably impossible to do moral reasoning in a total theoretical vacuum where we never draw on our pre-theoretical judgments. Thinking through our real-world moral experiences is a big part of how we come to learn and understand what we morally ought to do. What we must remember is that some intuitive judgments will not survive critical reflection: they will prove inconsistent with other judgments we make or with deeply plausible theoretical principles. We must avoid dogmatic commitment to our intuitive judgments and be willing to revise or discard them. The title of *considered judgment* should be reserved only for those intuitions that have survived sustained critical scrutiny and remained stable over time. When I appeal to an intuition about a case or scenario, my aim is to appeal to a considered judgment rather than a reflexive brute intuition.

How the book proceeds

This book is divided into four main parts. Part I, which contains this chapter and the next, provides an overview of the population problem and how I approach it. This chapter has established the background context and the underlying moral methodology. Chapter 2 focuses on the empirical facts that connect population growth to a variety of environmental problems. Understanding these facts is a prerequisite for determining how we should respond to the problem. I start by discussing skepticism about the gravity of the population problem, which is motivated in part by the inaccurate predictions made by Thomas Malthus and Paul Erlich about the effects of population growth. I then explain why this attitude is mistaken by highlighting some of the major features of the environmental problems that are currently unfolding.

Part II contains the text's central moral argument and explores its implications. In Chapters 3 and 4, I present and defend the Population Reduction Argument, which concludes that we have a moral duty to reduce our current population. In Chapter 3, I start with the moral duty to avoid causing massive and unnecessary harm and proceed to the conclusion that we should dramatically reduce our levels of environmental degradation. In Chapter 4, I argue that our obligation to dramatically reduce our levels of environmental degradation cannot be adequately met without trying to reduce population size. Since our population is still growing, we should aim to reduce fertility rates such that the global population peaks and then starts to shrink. I remain agnostic about what precise number of people we should aim for, but it is clear that our current number of 7.8 billion people is well above a population that is sustainable in the long term.

Chapters 5 and 6 address what follows if the conclusion of the central argument is accepted. Chapter 5 focuses on what policy measures we could permissibly pursue to lower fertility rates. I argue that the most effective measures for slowing (and eventually reversing) population growth do not involve significant coercion. In fact, many of them actually *increase* people's autonomy by making it easier for them to choose the number of children they want to have. In Chapter 6, the scope shifts from collective moral obligations to individual moral obligations. I examine the implications of population growth for individual procreative decision-making. Ultimately, for those who are able to do so, I argue that prospective parents should limit themselves to one biological child per person so that they do not reproduce above the replacement level fertility rate.

Once the Population Reduction Argument and its implications have been presented in Part II, I address some objections to my position in Part III. In Chapter 7, I examine concerns that my proposed measures to encourage lower rates of procreation are too lax. These objections are unified by the thought that procreation is, at least under present conditions, usually wrong. Those who hold this view would object that my policy measures do not go far enough in restricting people's procreative activity. In Chapter 8, I consider concerns that my proposed policies are too restrictive and not consistent with respecting people's procreative rights. I refute both these general objections.

After addressing these objections to the Population Reduction Argument, I turn my attention to a few lingering questions about the moral dimensions of population growth. In Chapter 9, I consider how the aim to reduce population growth could influence domestic immigration policies. Since immigration policy could be the topic of another book entirely, I limit my scope to two important connections. First, I consider how the imperative to protect the welfare of future people could motivate nations with large ecological footprints to restrict immigration. Doing so would make it easier for these countries to reduce their environmental impacts since population growth due to immigration would be lower. Second, I assess the extent to which immigrants could provide a solution to the short-term economic challenges associated with low fertility rates. Allowing more immigration to increase the size of the working class could alleviate the burdens on the younger members of a nation as its population ages. These two considerations do not answer the larger question of what a just immigration policy would look like, but they do illustrate two ways in which concerns about population growth should factor into that broader discussion.

Chapter 10 expands the scope of moral concern to the nonhuman community. The Population Reduction Argument is an anthropocentric one – it focuses only on the interests, needs, and values of human beings. Thus, the effects on animals and the environment are only considered insofar as they matter to human beings. In this chapter, I briefly assess how extending some moral consideration to nonhuman animals and the environment would affect the implications of my argument. The answer ultimately proves rather straightforward: the greater weight we place on the moral standing of those in the nonhuman world, the *stronger* the imperative to reduce population becomes.

The book concludes in Chapter 11 with a short reflection on whether we can solve the environmental problems on the horizon. I highlight a flaw in assessing the problems in this manner: they are not the kind of phenomena that have full-fledged solutions in the way that some other moral problems do. Hence, I insist that we instead ask the question of how we can make the future better and consider how to answer that question as the rest of the century unfolds.

This text is also affixed with an appendix that addresses the non-identity problem. This problem in intergenerational ethics refers to the observation that the identities of some future people are dependent on the actions that we take in the present. Different courses of action will cause different people to meet or existing couples to procreate at different times, resulting in entirely different people being born. Since these people would not exist if we did not act in a particular way, some believe that they cannot be harmed by these actions even when the actions appear to adversely affect their welfare. I address the non-identity problem briefly in Chapter 3, but since it has occupied a prominent place in the literature on population ethics, I have assembled this appendix for those who want a more thorough assessment of it.

Notes

1 Unless otherwise specified, I use the term "we" to refer to all human beings who presently exist.
2 This is not an ironclad rule: some countries with extremely large populations (such as China and India) have larger carbon footprints than some developed nations even though their per capita emissions are still relatively low. The more general point is simply that industrialized wealthy nations are typically considered more responsible for climate change due to their present and past contribution to the problem.
3 As one illustration, scientists in the United States appear significantly more worried about population growth than the general public. In a survey conducted by the Pew Research Center, 59 percent of the general public stated that there will not be enough food and resources to distribute around the globe if population growth continues. Among members of the American Association for the Advancement of Science, 82 percent held this position (Funk et al. 2015, p. 51).
4 The effect of plagues and epidemics on population size is usually transitory, but that did not stop people from believing that Africa's population size would see a serious decline. We now know that this prediction was grossly mistaken: many African countries have extremely high population growth rates compared to the rest of the world.
5 Sufficient expansion may have been impossible even without eliminating population from the discussion, but the point is that limiting discussion of population likely hindered this expansion rather than helping it.
6 In some parts of the world, the effects of population growth are more pronounced – especially in areas where wilderness is rapidly disappearing. But many people are insulated from natural environments and unaware of the research

on population growth. These individuals are unlikely to recognize how pervasive or significant its effects really are.
7 I borrow this example from Palmer (2012, pp. 98–99).
8 In their survey of professional philosophers, Bourget and Chalmers (2014) found that respondents "accepted" or "leaned toward" the major positions in normative ethics with the following frequency: deontology, 25.9 percent; consequentialism, 23.6 percent; virtue theory, 18.2 percent; and other, 32.3 percent. Utilitarianism is just one type of consequentialist theory, so clearly an argument that depends on accepting that view will only appeal to a small range of philosophers.
9 For one of the most recent and ambitious attempts at developing a unified field theory of ethics, see Parfit (2011).
10 Nelson Goodman (1955, pp. 65–68) appears to be the first philosopher to explicitly describe and endorse this method, though Goodman employed it as a means of justifying principles of deductive and inductive inferences. John Rawls (1999, pp. 18–19, 42–45) is responsible for popularizing the term.

References

Bourget, David, and David Chalmers. 2014. "What Do Philosophers Believe?" *Philosophical Studies* 170: 465–500.

Bricker, Darrell, and John Ibbitson. 2019. *Empty Planet: The Shock of Global Population Decline*. London: Robinson.

Campbell, Martha. 2012. "Why the Silence on Population?" In *Life on the Brink: Philosophers Confront Population*, eds. Phil Cafaro and Eileen Crist, 41–55. Athens: University of Georgia Press.

Clarke, Adele, and Donna Haraway (eds.). 2018. *Making Kin Not Population*. Chicago: Prickly Paradigm Press.

Conly, Sarah. 2016. *One Child: Do We Have a Right to Have More?* Oxford: Oxford University Press.

Coole, Diana. 2018. *Should We Control World Population?* Cambridge: Polity Press.

Crist, Eileen. 2019. *Abundant Earth: Toward an Ecological Civilization*. Chicago: University of Chicago Press.

Foreman, Dave. 2012. "The Great Backtrack." In *Life on the Brink: Philosophers Confront Population*, eds. Phil Cafaro and Eileen Crist, 56–71. Athens, GA: University of Georgia Press.

Funk, Cary, Lee Rainie, Aaron Smith, Kenneth Olmstead, Maeve Duggan, and Dana Page. 2015. "Public and Scientists' Views on Science and Society." www.pewresearch.org/internet/wp-content/uploads/sites/9/2015/01/PI_ScienceandSociety_Report_012915.pdf. Accessed November 22, 2019.

Goodman, Nelson. 1955. *Fact, Fiction, and Forecast*. Cambridge, MA: Harvard University Press.

Greene, Joshua, R. Brian Sommerville, Leigh Nystrom, John Darley, and Jonathan Cohen. 2001. "An fMRI Investigation of Emotional Engagement in Moral Judgment." *Science* 293: 2105–2108.

Haidt, Jonathan. 2001. "The Emotional Dog and Its Rational Tail: A Social Intuitionist Approach to Moral Judgment." *Psychological Review* 108, no. 4: 814–834.

Jehl, Douglas. 2002. "Saving Water, U.S. Farmers Are Worried They'll Parch." *New York Times*, August 28, p. 1.

Lutz, Wolfgang, Warren Sanderson, and Sergei Scherbov. 2001. "The End of World Population Growth." *Nature* 412: 543–545.

Palmer, Tim. 2012. "Beyond Futility." In *Life on the Brink: Philosophers Confront Population*, eds. Phil Cafaro and Eileen Crist, 98–107. Athens, GA: University of Georgia Press.

Parfit, Derek. 1982. "Future Generations, Further Problems." *Philosophy & Public Affairs* 11, no. 2: 113–172.

Parfit, Derek. 2011. *On What Matters: Volume One*. Oxford: Oxford University Press.

Pearce, Fred. 2010. *The Coming Population Crash and Our Planet's Surprising Future*. Boston: Beacon Press.

Potts, Malcolm, and Martha Campbell. 2005. "Reverse Gear: Cairo's Dependence on a Disappearing Paradigm." *Journal of Reproduction & Contraception* 16, no. 3: 179–186.

Rawls, John. 1999. *A Theory of Justice: Revised Edition*. Cambridge, MA: Belknap Press.

Rieder, Travis. 2016. *Toward a Small Family Ethic: How Overpopulation and Climate Change Are Affecting the Morality of Procreation*. Cham, Switzerland: Springer.

Roberts, David. 2018. "I'm an environmental journalist, but I never write about overpopulation. Here's why." *Vox*. www.vox.com/energy-and-environment/2017/9/26/16356524/the-population-question. Accessed 9/22/2019.

Shragg, Karen. 2015. *Move Upstream: A Call to Solve Overpopulation*. Minneapolis–St. Paul, MN: Freethought House.

Simon, Julian L. 1996. *The Ultimate Resource 2*. Princeton, NJ: Princeton University Press.

Smail, J. Kennth. 1997. "Beyond Population Stabilization: The Case for Dramatically Reducing Global Human Numbers." *Politics and the Life Sciences* 16, no. 2: 183–192.

2 The gravity of the population problem

Before we engage with the ethical implications of population growth, we have to assess how significant the problem is. As an initial observation, a large population is not automatically a bad thing. It is not as if adding an additional person to the world automatically means that someone else in the world must live a worse life, and there can certainly be situations where increasing population size can be good for a society.[1] Under our current circumstances, the main problem with population growth is that it multiplies the negative impacts associated with common behaviors (Ryerson 2010). Three billion people polluting is bad, *four* billion people polluting is worse, and *five* billion people polluting is worse still. Our actual numbers are much higher than five billion. At the time of writing, Earth contains about 7.8 billion human beings. By the time you read this sentence, that number might be much higher.[2] The UN Department of Economic and Social Affairs (2019) estimates that the global population will be nearly 11 billion people in 2100.[3]

In broad terms, population growth is a problem because it places strain on our planet's resources. Earth has a limited quantity of farmable land, water, carbon sinks, or inhabitable space where people can live comfortably. Thus, if population rises indefinitely, we will eventually overshoot our ecological limits and will not be able to support our numbers. Of course, since fertility rates around the world are generally declining, the threat is not that population growth will persist forever. Rather, the worry is that we will cross ecological boundaries before population growth ceases and that many people will suffer or die as a result.

If this concern sounds familiar, that is because this concern has been voiced in the past by Thomas Malthus (2007 [1798])) and Paul Erlich (1975). Writing more than two centuries ago, Malthus claimed that population growth would outstrip food supply and soon lead to widespread starvation. He was obviously wrong about that.[4] Paul Erlich made a similar prediction in 1968 when he originally published *The Population Bomb*. He suggested that we would experience widespread starvation in the 1970s and 1980s if population growth continued, but this dire outcome did not come to pass.[5]

Given the inaccuracy of Malthus and Erlich's predictions, it is tempting to conclude that concerns about population are destined to be exaggerated.

These recent lessons from our history appear to illustrate that technological progress can enable us to accommodate our rising numbers. But reaching this conclusion on the basis of two examples is far too hasty, especially since there are some significant differences between our current circumstances and the scenarios that Malthus and Erlich envisioned. For one, Malthus and Erlich were focused on the available food supply, but the impacts of population size are not limited to food availability. For another, Malthus and Erlich were speculating about the impacts of population growth. With respect to most of the problems that concern us now, we do not have to guess about what might take place because the impacts are already observable.

Nevertheless, there is one important takeaway from Malthus and Erlich's inaccurate predictions: the Earth does not have a fixed carrying capacity. Technological improvements, such as those in agriculture, can increase the number of people that can inhabit the Earth. Simultaneously, strains on available resources, such as soil erosion (Pimentel 2006) or overfishing (UN Food and Agricultural Organization 2016), can decrease the number of people who can inhabit the Earth.[6] Even though Earth's carrying capacity can change, some have argued that the long-term sustainable limit for global population is about two billion (Crist 2019, p. 201; Foreman 2012; Pimentel et al. 2010; Smail 1997). Karen Shragg (2015) sets the number even lower – 1.5 billion (p. 94). The malleability of Earth's carrying capacity raises doubts about our ability to define this limit, but even if they underestimate the Earth's long-term carrying capacity by a few billion people (which would be an extraordinary underestimate), we would still be overshooting this threshold by billions. Moreover, when we examine the list of mounting problems that intertwine with our population size, it is not difficult to identify ways in which we are pushing ecological limits.

While we are still currently able to produce enough food globally to feed everyone (if this food were distributed more equitably), one estimate from the UN Food and Agriculture Organization (2006) suggests that the necessary supply of food will increase dramatically – by up to 70 percent by 2050. Additionally, a recent update to this report did not change these estimates, although the authors do express a bit more pessimism in the ability to meet the global demand for food in the future (Alexandratos and Bruinsma 2012, pp. 37–38). Moreover, it is not realistic to imagine a world with an ideal and equitable distribution of food: the regional differences in population size and farmable land, especially when combined with national differences in political and economic circumstances, render this outcome a practical impossibility. There are almost 800 million people in the world who are undernourished (UN Food and Agricultural Organization 2015), and as population grows, it will only become more difficult to feed those in need.

Water shortages are another serious concern. Groundwater plays a crucial role in irrigating crops, providing water to those who need it, and maintaining the health of local ecosystems (Giordano 2009; Siebert et al. 2010). Groundwater is a renewable resource, but it can still be depleted when our rate of consumption exceeds the rate at which it replenishes. Our current practices are depleting groundwater at 3.5 times the sustainable rate, leaving 1.7 billion people living in areas where their groundwater resources or the ecosystems that depend on groundwater (or both) are threatened (Gleeson et al. 2012). A greater population will make this overconsumption of groundwater even more difficult to reverse.

Shortage of resources often leads to conflicts between or within nations, and we know all too well that this can result in war. Water shortage played a significant role in the recent civil war in Syria (Gleick 2014).[7] A lack of available land and inequity in land distribution contributed significantly to the civil war that began in Rwanda in 1994 (André and Platteau 1998). As population rises, we can expect these types of conflicts to arise more frequently.[8]

These concerns alone are enough to justify taking population growth seriously, but they are arguably not the most significant problems to which population growth contributes. Climate change has emerged as one of this century's biggest moral challenges, and biodiversity loss has become so pronounced that some fear we have initiated the world's sixth mass extinction event (Ceballos et al. 2015). Given the scope and magnitude of these impacts, each of them warrants a robust explanation.

Climate change

The changes in the global climate system that we are now observing result from people emitting greenhouse gases (GHGs) like carbon dioxide, methane, ozone, and nitrous oxide. These gases absorb infrared radiation from sunlight, thereby trapping it in the atmosphere for a period of time. During the last two centuries, increased emissions of GHGs have caused the average global temperature to rise significantly. Between 1880 and 2012, average global surface temperature increased by about 0.85°C (IPCC 2014b, p. 2).[9] That may not sound like a significant increase, but the average global temperature during the most recent ice age was only 5°C lower that the average global preindustrial temperature. Relatively small changes in global temperature can have enormous impacts.

The majority of the increase in global average temperature is a result of our emissions of carbon dioxide into the atmosphere. Pre-industrial levels of carbon dioxide were about 275 parts per million (ppm) by volume in the atmosphere. We have now surpassed 400 ppm of carbon dioxide in the atmosphere (Kahn 2016).[10] The effects of climate change on other human beings are significant and widespread. I will highlight some of the major effects, all of which are summarized in recent reports from the IPCC (2014a, 2014b).

Perhaps the most powerful way to understand the harm of climate change is to consider the number of deaths it will cause. Reducing the impacts of climate change to monetary values can obscure the grim realities of the issue – namely, that large numbers of people will die or suffer severely as climate change continues (Nolt 2015). One study from the World Health

Organization (2005) concludes that climate change may have been responsible for at least 150,000 deaths in 2003. One of their later reports (World Health Organization 2009) reaches a similar conclusion: by 2004, the annual global death toll from climate change had reached 140,000 people. Figures from the Global Humanitarian Forum (2009) suggest that these estimates are too low: their research estimates that 300,000 people die from climate change annually with the majority of those deaths occurring in developing nations. More recent estimates are even bleaker. DARA (2012) suggests that the annual death toll from climate change is about 400,000. Their research, like the study conducted by the Global Humanitarian Forum, indicates that most of those deaths take place (and will continue to take place) in developing nations. They also project that the annual death toll from climate change could reach 700,000 by 2030.[11] In another recent study, Marco Springmann and his colleagues (2016) estimate that 529,000 annual deaths are being caused by climate change due to its effects on agriculture and food security.

While there is some variance among these casualty counts, two facts are clear. First, on any plausible estimate, hundreds of thousands of people are already dying annually from climate change. As John Broome (2012, p. 33) notes, those casualty rates indicate that tens of millions of deaths will be caused this century even if the annual casualty counts do not increase. But that brings us to the second obvious conclusion from these studies: the number of annual deaths from climate change will almost surely increase as the effects of climate change become more severe. By the end of the century, we could see a death toll that is much higher than just tens of millions of people.

However, we should not focus only on the impacts that will occur in the next 80 years. Another crucial feature of climate change is its long-lasting nature. The temperature increase resulting from climate change will, unless we perform extraordinary feats of geoengineering, persist for tens of thousands of years, if not longer (Archer et al. 2009). According to some models, changes to surface temperatures will persist for 23,000 to 165,000 years (Zeebe 2013). If we tie this observation with the estimated death tolls caused by climate change each year, we immediately reach a stark conclusion: climate change could lead to *billions* of deaths over the next millennium, depending on its severity and the extent to which we are able to adapt.[12]

We also cannot overlook the fact that many affected by climate change will not die but will nevertheless suffer significantly. Climate change will increase the prevalence of severe weather events, such as droughts, heatwaves, and hurricanes. Increased surface temperatures will make agriculture more difficult in certain parts of the world, and ocean acidification will reduce the food productivity of the oceans (IAP 2009). Temperature increases will alter and expand the range in which many insects can survive, causing more people to become vulnerable to various diseases they carry. These effects can cause death, of course, but more often they result in suffering. People survive but have their quality of life reduced, often severely – at least in the case of dehydration, malnourishment, and sickness. Although it is difficult to estimate the number of people who will suffer significantly (but not fatally) from climate change with precision, the widespread distribution of its effects and their severity indicate that these numbers are massive – in all likelihood at least comparable to the number of annual deaths caused by climate change.

Military leaders are also concerned about climate change threatening national security by creating mass migrations (Carrington 2016). Droughts and other resource shortages caused by heat waves and other severe weather events may destabilize regions and lead to war. Sea level rise will cause islands to disappear and coastlines to gradually creep inland. In fact, some of those impacts are already observable. Five of the Solomon Islands vanished beneath the sea between 1947 and 2014, and six other islands in this area are experiencing severe shoreline recession (Albert et al. 2016). Some coastal cities in the United States are flooding more frequently as a result of climate change (Gillis 2016). In extreme cases, inhabitants of these regions have no choice but to relocate. The result is that millions of people will be displaced, which raises the possibility of a mass migration crisis.

Finally, climate change increases the rate of species extinctions (Thomas et al. 2004). As ocean and surface temperatures increase, species often become unable to survive in the niches that they inhabit. They migrate toward the poles or to higher elevations when possible, but many species are simply unable to adapt to their rapidly changing environments and go extinct. As one prominent example, coral reefs can be utterly decimated by climate change because they are stationary and cannot escape warming ocean water (Hughes et al. 2018). Biodiversity loss is not just caused by climate change, though, and the ongoing disappearance of species is severe enough that the topic deserves its own separate treatment.

Biodiversity loss

I use the general term "biodiversity" to refer to global species diversity. Thus, biodiversity loss refers to decrease in global species diversity caused by human action. Human-caused biodiversity loss has been occurring for some time, and climate change is just one of its many causes (Barnosky et al. 2011).[13] But while we can readily assert that biodiversity is in decline, determining the extent of the decline is more challenging (Wilson 2016, ch. 3). The main difficulty lies in determining what the pre-human rate of extinction is – that is, the rate at which species would go extinct if not for the impact of human beings. Without an accurate estimate of that rate, we cannot know how much our actions are increasing the rate at which extinctions would ordinarily occur. Obtaining accurate numbers on how many species exist and how many are being lost is similarly difficult. New species are discovered every year, and extinctions often go unnoticed because the lost species are unknown to us. Even so, existing estimates are utterly disheartening. Excluding bacteria, there are an estimated 8.7 million species on Earth (Mora et al. 2011). While conservation biologists previously estimated the pre-human extinction rate at about one species per million per year, recent studies suggest that this figure is actually about 0.1 species per million per year (De Vos et al. 2015; Pimm et al. 2014). That means that the current estimated rate of species extinctions – roughly 100 per million per year – is an astonishing 1000 times the rate at which extinctions would occur without the impact of human actions!

The main contributors to biodiversity loss are captured in the acronym HIPPO (Wilson 2016, pp. 57–58): habitat destruction, invasive species, pollution, population growth, and overhunting (including overfishing). While population growth is listed as its own factor, it also contributes to all of the other items of this list. More people means a greater need for space and resources, and this often leads to habitat destruction when land is cleared for housing or farming. More people means a greater need for food, which can cause regional overfishing as demand for fish increases. More people means more traveling and a greater need to transport goods across borders. Non-native species often get transported to new environments unintentionally, and sometimes, they can eliminate native species by overtaking their ecological niche.[14]

Naturally, the rapid loss of biodiversity has a lot of conservation biologists concerned. Biodiversity loss affects human beings in many ways. Perhaps most significantly, human beings are affected by reductions in biodiversity by being deprived of the ecosystem services that biodiversity enables. Ecosystem services refer to the "properties of ecosystems that either directly or indirectly benefit human endeavors, such as maintaining hydrologic cycles, regulating climate, cleansing air and water, maintaining atmospheric composition, pollination, soil genesis, and storing and cycling of nutrients" (Hooper et al. 2005, p. 7). These services provide the basic conditions needed for people to survive, and they would be costly to provide by other means (if it could even be done).

Biodiversity is also a source of much joy for people, whether it stems from aesthetic appreciation of exotic species (such as the peacock) or curious fascination with the most bizarre ones (such as the blob fish). Alan Carter (2010) even argues that the best reason to preserve biodiversity is rooted in aesthetics and that the loss of a species is analogous to the loss of an entire genre of music or film (pp. 73–75).[15] Certainly, there are many species that can be considered beautiful, such as the majestic bald eagle or the graceful antelope, but even species that strike us as outright hideous, such as the Amazonian giant centipede, can have robust aesthetic value. Aesthetic value is not limited to beauty alone. After all, well-made horror films have aesthetic value even if their tone, imagery, and subject matter are far from beautiful.

Biodiversity is also a bountiful source of knowledge. E. O. Wilson (1992) describes biodiversity as "The Great Encyclopedia of Life" – a relatively untapped font of knowledge that "would occupy 60 meters of library shelf per million species" even if each species occupied only a page in the volume (p. 151). Our scientific understanding of the nonhuman world is vastly incomplete, and the ability to study other life forms can offer crucial insights into how nonhuman life forms interact. Beyond fulfilling the scientific interests of many people, these discoveries can also give us insight into ways in which we might improve the welfare of human beings. For example, biodiversity provides a source of biologically active compounds that can aid the development of medicines crucial for promoting human health (Butkus 2015).

Many further reasons for valuing biodiversity could be offered.[16] Nevertheless, not everyone really believes that biodiversity loss is important or that its loss is particularly bad. One of the most robust critiques of biodiversity's value comes from Don Maier (2012). He surveys 12 different reasons that one might value biodiversity and argues that none of them can sufficiently ground biodiversity's value. His basic strategy is to propose a particular reason as the core of biodiversity's value and then raise counterexamples to this proposal. In the case of ecosystem services, for instance, Maier (2012) notes that the discussion of these services often omits discussion of ecosystem disservices (p. 167) and then lists many examples of ways in which the services that biodiversity provides are not always so beneficial. When evaluating the claim that biodiversity is a source of valuable medicines and pharmaceuticals, Maier (2012) mentions that only a small portion of all the existing species actually provide this benefit (pp. 196–206) and that biodiversity can often *increase* the incidence of disease (pp. 207–220). After he finishes surveying a purported reason to value biodiversity, he concludes that it cannot be the core of biodiversity's value and moves onto the next reason. After finding problems with all 12 proposed reasons, he concludes that biodiversity must not be particularly valuable.

Defenders of biodiversity's value can respond to Maier in several ways. They may point out that Maier does not address nearly all the reasons on offer for thinking biodiversity is valuable. In other places, they may simply deny the plausibility of his arguments. For example, his reasons for thinking that biodiversity's value cannot be primarily epistemic is that there would be much to learn "from a vastly changed biological world that contained a significantly different set of species with significantly different population sizes (abundances)" and that "the very processes involved in bringing about such an altered world ... would be a rich source of knowledge that could not be tapped except by observing them unfold" (Maier 2012, p. 235). Certainly, there might be some knowledge to gain from such events, but it is difficult to believe that this knowledge would actually be comparable to what we can learn by studying the vast array of different species around the world, especially since so many of them have yet to even be discovered. It is also doubtful that the knowledge obtained by facilitating a mass extinction would be as instrumentally valuable as greater knowledge of currently existing species. Studying currently extant species could yield new insight into ways that these species contribute (or could contribute) to human flourishing. Studying their demise might give us similar insight, but the instrumental value of that knowledge would be rather low because the species would be gone and unable to make those contributions.

I lack the space to survey every individual argument Maier makes, so instead I will highlight a flaw that undermines his general argumentative strategy. The central problem is that Maier's argument is invalid: the truth of his argument's premises does not entail the truth of his conclusion. He assumes that because none of the individual reasons he considers can be the core of biodiversity's value, it is not valuable. But that does not follow: some things are valuable for a plurality of reasons that cannot be reduced to a single core value. Take love as an example.

If we were trying to determine what makes love valuable, we might initially answer that love makes people happy, and this claim is often true: many people report being their happiest when they are in loving relationship with someone. But love can also be a source of great sorrow when our love is not reciprocated or when our loved one suffers or dies. Thus, the core value of love cannot just be that it makes us happier. Perhaps we think love is a means of developing virtuous character traits like sympathy, empathy, and kindness. Certainly, it is not hard to see why loving someone will acquaint a person with these virtues, and yet, love can also serve as motivation for many despicable deeds when concern for one's beloved trumps other moral considerations. Thus, the core of love's value cannot be in the cultivation of moral virtue. Perhaps the phenomenon of loving someone is unique and provides special epistemic insights into human nature. There is little doubt that deeply loving someone profoundly changes one's outlook on the world, but simultaneously, we pursue love even after we have experienced it many times. There is also much we can learn from the loss of love. The unique knowledge that love often provides cannot be the core of its value.

We could repeat this process with love many more times without identifying any central reason why love is valuable. But notice that it does not follow from this result that love has no value. Aside from the most nihilistic among us, people tend to agree that love is valuable. Love has many sources of value, and what makes it valuable in a particular context can vary. We can reach the same conclusion about biodiversity's value: what makes it valuable varies depending on the context and cannot be reduced to a single source.

This result should not be that surprising because not all valuable things in the world have a single source of value. Beyond love, we might add items like friendship, beauty, moral virtue, wilderness, and sex. There is nothing strange in understanding the value of biodiversity in the same way. Certainly, in some particular cases, its value will not originate in ecosystem services or its means for creating new medicines. Some species may not ultimately prove to be treasure troves of knowledge or to add great aesthetic value to the world. But how often is it going to be the case that *none* of the many reasons for thinking that biodiversity is valuable will apply? How frequently could we run through every single item on the list and deem them all inapplicable? Much as love rarely fails to add something of great value to our lives, biodiversity rarely fails to contribute something of great value to the world. Thus, Maier's skepticism about the value of biodiversity is unwarranted.

It is also worth acknowledging one of the most significant reasons that scientists tend to care about biodiversity: its elimination could lead to ecosystem collapse. Maier may be skeptical about the relationship between biodiversity and the maintenance of ecosystem services, but robust meta-analyses demonstrate a general consensus that biodiversity correlates positively with ecosystem functioning (Balvanera et al. 2006; Cardinale et al. 2012). What this means is that greater biodiversity is generally associated with greater ecosystem functioning and lower biodiversity is generally associated with poorer ecosystem functioning. The relations between species and the role that each plays in maintaining ecosystem functioning will vary significantly across different ecosystems, but a substantial decrease in the number of species in an ecosystem is very likely to have an adverse effect on ecosystem functioning. We are on pace to eliminate between one-quarter and two-thirds of all currently existing species (Myers 1993; Myers and Knoll 2001).

Since the loss of species will be so substantial, it is reasonable to be worried about the destruction of ecosystems and the elimination of the ecosystem services that they provide. This is one of the main reasons why E. O. Wilson (2016) claims that species extinctions rank "with pandemics, world war, and climate change as among the deadliest threats that humanity has imposed on itself" (Wilson 2016, p. 187). Because Wilson views the problem as so severe, he advocates setting aside half the Earth for the preservation of wildlife. Paul Watson (2012), comparing Earth to a spaceship, describes the problem as follows:

> Biodiversity destruction is the single greatest threat to human survival on this planet because it weakens and removes our custodians, the species that make it possible for us to be the passengers. What we are in effect doing is eroding the immune system of the planet, compromising the functioning of Earth's life-support system. We have become like a deadly autoimmune disease to Earth, killing the essential crewmembers as we overload our spaceship with human passengers.
>
> (p. 132)

Wilson and Watson's grave language may strike some as hyperbolic. Given humanity's resourcefulness and our prevalence across the globe, it seems doubtful that biodiversity loss could lead to our own extinction. What is more likely is that massive biodiversity loss will manifest in a variety of less extreme effects: economic costs from ecosystem services that disappear, a reduction in the availability of pharmaceuticals, greater difficulties providing vital resources in certain regions, the permanent loss of the knowledge and beauty associated with particular species, and so on. In more practical terms, these effects will not lead to our extinction but to people living in a bleaker world – one where resources are scarcer, our numbers are higher, and the general welfare of humanity is lower.

Responding to our situation

In the near future, we face the prospect of regional food and water shortages, significant threats to human health, widespread and rapid species extinctions, the displacement of hundreds of millions of people, and a drastic rise in the risk of war. Even based only on the limited survey I have undertaken here,[17] these problems can make the future look grim. The Science and Security Board of the *Bulletin of the Atomic Scientists* (2020) have placed the Doomsday Clock at 100 seconds to midnight, citing climate change as one of the two major perils that place us on the precipice of global catastrophe.[18] A recent article in *BioScience* that encapsulates many scientists' worries about the state of the planet gathered more than 15,000 signatures of support (Ripple et al. 2017). Those familiar with the environmental degradation around the world are doing their best to make the significance of the problem known. Now we must consider what to do in response.

Our current situation is the result of large numbers of people engaging in environmentally destructive behaviors. Overwhelmingly, the focus of the past three decades has been on changing our patterns of consumption so that each person has a smaller impact on the environment. Generally, the possibility of decreasing our environmental impact by lowering our numbers has been treated as an afterthought or omitted from discussion altogether. In the next part of the book, I argue that this trend has pushed us in the wrong direction: a significant part of our response should be attempting to stabilize global population and then gradually reduce it.[19]

Notes

1. As an extreme illustration, imagine that our numbers dwindled so gravely we were faced with the serious threat of human extinction. Under those conditions, increasing the population size would surely be favored over further reduction or the maintenance of perilously low numbers.
2. Trackers monitoring global population in real time are easily accessible online. For one such source, see Worldometers (2019).
3. There is a range of possible outcomes, but they judge this number, which originates from the medium variant of their projections, as the most likely outcome. While this is the most comprehensive population data available, not everyone agrees that global population will actually peak at such a high number. See Bricker and Ibbitson (2019) for a recent critique of the UN's findings.
4. Beyond getting his empirical claims wrong at the time, Malthus was also criticized by Karl Marx and Friedrich Engels. For an overview of their criticisms, see Charbit (2009, ch. 5).
5. Erlich presented hypothetical scenarios that illustrated dire outcomes caused by rising population. Although he said that these scenarios were "just possibilities"

and "not predictions" (Erlich 1975, p. 49), their presentation in combination with the alarmist tone of his book created the impression that he viewed these scenarios as realistic.
6 Perhaps this explains why Joel Cohen (1995) never concretely answers the titular question of his *How Many People Can the Earth Support?*
7 While this drought did take place and play some role in the ensuing war, the scale of the drought and the extent to which climate change caused it are both controversial matters. See Kelley et al. (2015) and Selby et al. (2017).
8 My focus in this chapter is one the broader effects of population growth – those felt at the national and global levels. Even so, it is worth noting that some of the adverse effects of population growth are more localized. In urban areas, for example, population growth can cause overcrowding, which has a variety of negative effects on human health (Gray 2001).
9 A rise of 0.85 °C is the most likely value. According to the IPCC's estimates, there is a 90 percent likelihood of the average warming having a value of between 0.65 and 1.06 °C during that time period.
10 To view the current levels of CO_2, gathered by the National Oceanic and Atmospheric Administration, see NASA (2019).
11 The World Health Organization (2014) estimates 250,000 annual deaths in 2030–2050 from climate change, but their estimate is not comprehensive because it does not account for certain "major pathways of potential health impact, such as the effects of economic damage, major heatwave events, river flooding and water scarcity" (p. 1). An additional survey on the effects of climate change on human health can be found in Kim, Kabir, and Jahan (2014).
12 Suppose that the average annual deaths caused by climate change for the next 1000 years is 200,000 – a relatively low estimate given the studies I have cited. This would still translate to two billion deaths (1000 × 200,000) caused by climate change over the next millennium.
13 For specific measures pertaining to vertebrate and mammal species, see Ceballos, Ehrlich, and Dirzo (2017). For some measure of the decline in the insect population, see Hallman et al. (2017).
14 Further details on the ways in which overpopulation contributes to biodiversity loss can be found in Foreman (2014, ch. 4).
15 Darrel Moellendorf (2014) also argues that a central aspect of biodiversity's value is its aesthetic value (ch. 2).
16 For example, biodiversity may have value because of its ability to transform our values and alter our preferences (Norton 1987; Sarkar 2005) or because of its maintenance is necessary to secure future people's autonomy (Zwarthoed 2016). Additionally, Rolston (1988, ch. 1) lists 14 different reasons for why human beings typically value nature, many of which can apply to why we should value biodiversity.
17 For a more thorough survey of the effects of population growth, see Weisman (2013).
18 The other peril is the potential use of nuclear weapons.
19 Significant portions of this chapter are derived from chapter 4 of my doctoral dissertation. See Hedberg (2017).

References

Albert, Simon, Javier Leon, Alistair Grinham, John Church, Badin Gibbes, and Colin Woodroffe. 2016. "Interactions Between Sea-Level Rise and Wave Exposure on

Reef Island Dynamics in the Solomon Islands." *Environmental Research Letters* 11: 054011. DOI: 10.1088/1748-9326/11/5/054011.

Alexandratos, Nikos, and Jelle Bruinsma. 2012. *World Agriculture towards 2030/2050: The 2012 Revision.* Rome: UN Food and Agricultural Organization. www.fao.org/docrep/016/ap106e/ap106e.pdf. Accessed December 6, 2019.

André, Catherine, and Jean-Philippe Platteau. 1998. "Land Relations under Unbearable Stress: Rwanda caught in the Malthusian Trap." *Journal of Economic Behavior & Organization* 34, no. 1: 1–47.

Archer, David, Michael Eby, Bictor Brovkin, Andy Ridgwell, Long Cao, Uwe Mikolajewicz, Ken Caldeira, Katsumi Matsumoto, Guy Munhoven, Alvaro Montenegro, and Kathy Tokos. 2009. "Atmospheric Lifetime of Fossil Fuel Carbon Dioxide." *Annual Review of Earth and Planetary Sciences* 37: 117–134. DOI: 10.1146/annurev.earth.031208.100206.

Balvanera, Patricia, Andrea Pfisterer, Nina Buchmann, Jing-Shen He, Tohru Nakashizuka, David Raffaelli, and Bernhard Schmid. 2006. "Quantifying the Evidence for Biodiversity Effects on Ecosystem Functioning and Services." *Ecology Letters* 9, no. 10: 1146–1156.

Barnosky, Anthony, Nicholas Matzke, Susumu Tomiya, Guinevere Wogan, Brian Swartz, Tiago Quental, Charles Marshall, Jenny McGuire, Emily Lindsey, Kaitlin Maguire, Ben Mersey, and Elizabeth Ferrer. 2011. "Has the Earth's Sixth Mass Extinction Already Arrived?" *Nature* 471: 51–57.

Bricker, Darrell, and John Ibbitson. 2019. *Empty Planet: The Shock of Global Population Decline.* New York: Robinson.

Broome, John. 2012. *Climate Matters: Ethics in a Warming World.* New York: W.W. Norton & Company, Inc.

Butkus, Matthew. 2015. "All Health is Local: Biodiversity, Ethics, and Human Health." *Ethics, Policy & Environment* 18, no. 1: 1–15.

Cardinale, Bradley, Emmett Duffy, Andrew Gonzalez, David Hooper, Charles Perrings, Patrick Venail, Anita Narwani, Georgina Mace, David Tilman, David Wardle, Ann Kinzig, Gretchen Daily, Michel Loreau, James Grace, Anne Larigauderie, Diane Srivastava, and Shahid Naeem. 2012. "Biodiversity Loss and Its Impact on Humanity." *Nature* 486, no. 7401: 59–67.

Carrington, Damian. 2016. "Climate Change Will Stir 'Unimaginable' Refugee Crisis, Says Military." *Guardian.*

Carter, Alan. 2010. "Biodiversity and All That Jazz." *Philosophy and Phenomenological Research* 80, no. 1: 58–75.

Ceballos, Gerardo, Paul Erlich, and Rodolfo Dirzo. 2017. "Biological annihilation via the ongoing sixth mass extinction signaled by vertebrate population losses and declines." *Proceedings of the National Academy of Sciences* 114, no. 30: E6089-E6096. DOI: 10.1073/pnas.1704949114.

Ceballos, Gerardo, Paul Erlich, Anthony Barnosky, Andrés Garcia, Robert Pringle, and Todd Palmer. 2015. "Accelerated Modern Human-Induced Species Losses: Entering the Sixth Mass Extinction." *Science Advances* 1, no. 5: e1400253. DOI: 10.1126/sciadv.1400253.

Charbit, Yves. 2009. *Economic, Social and Demographic Thought in the XIXth Century: The Population Debate from Malthus to Marx.* Dordrecht, Netherlands: Springer.

Cohen, Joel. 1995. *How Many People Can the Earth Support?* New York: W. W. Norton & Company.

Crist, Eileen. 2019. *Abundant Earth: Toward an Ecological Civilization.* Chicago: University of Chicago Press.

DARA (Development Assistance Research Associates). *Climate Vulnerability Monitor 2nd Edition: A Guide to the Cold Calculus of a Hot Planet.* Madrid: DARA and the Climate Vulnerable Forum, 2012. http://daraint.org/climate-vulnerability-monitor/climate-vulnerability-monitor-2012/report/. Accessed December 6, 2019.

De Vos, Jurriaan, Lucas Joppa, John Gittleman, Patrick Stephens, and Stuart Pimm. 2015. "Estimating the Normal Background Rate of Species Extinction." *Conservation Biology* 29, no. 2: 452–462.

Erlich, Paul. 1975. *The Population Bomb*, Revised ed. Jackson Heights, NY: Rivercity Press.

Foreman, Dave. 2012. "The Great Backtrack." In *Life on the Brink: Philosophers Confront Population*, eds. Phil Cafaro and Eileen Crist, 56–71. Athens, GA: University of Georgia Press.

Foreman, Dave. 2014. *Man Swarm: How Overpopulation Is Killing the Wild World*, 2nd ed., edited by Laura Carroll. LiveTrue Books.

Gillis, Justin. 2016. "Flooding of Coast, Caused by Global Warming, Has Already Begun." *New York Times.* www.nytimes.com/2016/09/04/science/flooding-of-coast-caused-by-global-warming-has-already-begun.html. Accessed December 6, 2019.

Giordano, Mark. 2009. "Global Groundwater? Issues and Solutions." *Annual Review of Environment and Resources* 34: 153–178.

Gleeson, Tom, Yoshihide Wada, Marc Bierkens, and Ludovicus van Beek. 2012. "Water Balance of Global Aquifers Revealed by Groundwater Footprint." *Nature* 488, no. 7410: 197–200.

Gleick, Peter. 2014. "Water, Drought, Climate Change, and Conflict in Syria." *Weather, Climate, and Society* 6, no. 3: 331–340.

Global Humanitarian Forum. 2009. *Climate Change: The Anatomy of a Silent Crisis.* www.ghf-ge.org/human-impact-report.pdf. Accessed December 6, 2019.

Gray, Alison. 2001. *Definitions of Crowding and the Effects of Crowding on Health: A Literature Review.* Wellington, NZ: Ministry of Social Policy. www.msd.govt.nz/documents/about-msd-and-our-work/publications-resources/archive/2001-definitionsofcrowding.pdf. Accessed December 6, 2019.

Hallman, Caspar, Martin Sorg, Eelke Jongejans, Hank Siepel, Nick Hofland, Heinz Schwan, Werner Stenmans, Andreas Müller, Hubert Sumser, Thomas Hörren, Dave Goulson, and Hans de Kroon. 2017. "More than 75 percent decline over 27 years in total flying insect biomass in protected areas." *PLoS ONE* 12, no 10: e0185809. DOI: 10.1371/journal.pone.0185809.

Hedberg, Trevor. 2017. "Population, Consumption, and Procreation: Ethical Implications for Humanity's Future." Ph.D. dissertation, Department of Philosophy, University of Tennessee.

Hooper, D. U., F. S. Chapin III, J. J. Ewel, A. Hector, P. Inchausti, S. Lavorel, J. H. Lawton, D. M. Lodge, M. Loreau, S. Naeem, B. Sachmid, H. Setälä, A. J. Symstad, J. Vandermeer, and D. A. Wardle. 2005. "Effects of Biodiversity on Ecosystem Functioning: A Consensus of Current Knowledge." *Ecological Monographs* 75, no. 1: 3–35. www.cedarcreek.umn.edu/biblio/fulltext/t2038.pdf. Accessed December 6, 2019.

Hughes, Terry, James Kerry, Andrew Baird, Sean Connolly, Andreas Dietzel, C. Mark Eakin, Scott Heron, Andrew Hoey, Mia Hoogenboom, Gang Liu, Michael McWilliam, Rachel Pears, Morgan Pratchett, William Skirving, Jessica Stella, and Gergely Torda. 2018. *Nature* 556: 492–496.

IAP (Interacademy Panel on International Issues). 2009. "IAP Statement on Ocean Acidification." www.interacademies.org/File.aspx?id=9075&v=462aaac6. Accessed February 18, 2020.

IPCC (Intergovernmental Panel on Climate Change). 2014a. *Climate Change 2014: Impacts, Adaptation, and Vulnerability*, edited by C. B. Field, et al. Cambridge: Cambridge University Press.

IPCC (Intergovernmental Panel on Climate Change). 2014b. "Summary for Policy Makers." *Climate Change 2014 Synthesis Report*. www.ipcc.ch/report/ar5/syr/. Accessed December 6, 2019.

Kahn, Brian. 2016. "Earth's CO2 Passes the 400 PPM Threshold – Maybe Permanently." *Scientific American*. www.scientificamerican.com/article/earth-s-co2-passes-the-400-ppm-threshold-maybe-permanently/. Accessed December 6, 2019.

Kelley, Colin, Shahrzad Mohtadi, Mark Cane, Richard Seager, and Yochanan Kushnir. 2015. "Climate Change in the Fertile Crescent and Implications of the Recent Syrian Drought." *Proceedings of the National Academy of Sciences* 112, no. 11: 3241–3246.

Kim, Ki-Hyun, Ehsanul Kabir, and Shamin Ara Jahan. 2014. "A Review of the Consequences of Global Climate Change on Human Health." *Journal of Environmental Science and Health*, Part C 32, no. 3: 299–318.

Maier, Donald. 2012. *What's So Good About Biodiversity? A Call for Better Reasoning About Nature's Value*. Dordrecht, Netherlands: Springer.

Malthus, Thomas. 2007 [1798]. *An Essay on the Principle of Population*. Meneola, NY: Dover Publications, Inc.

Moellendorf, Darrel. 2014. *The Moral Challenge of Dangerous Climate Change: Values, Poverty, and Policy*. New York: Cambridge University Press.

Mora, Camilo, Derek Tittensor, Sina Adl, Alastair Simpson, and Boris Worm. 2011. "How Many Species Are There on Earth and in the Ocean?" *PLOSBiology* 9, no. 8: e1001127. DOI: 10.1371/journal.pbio.1001127.

Myers, Norman. 1993. "Questions of Mass Extinction." *Biodiversity & Conservation* 2, no. 1: 2–17.

Myers, Norman, and Andrew Knoll. 2001. "The Biotic Crisis and the Future of Evolution." *Proceedings of the National Academy of Sciences* 98, no. 10: 5389–5392.

NASA (National Aeronautics and Space Administration). 2019. "Carbon Dioxide." *Global Climate Change: Vital Signs of the Planet*. https://climate.nasa.gov/vital-signs/carbon-dioxide/. Accessed December 6, 2019.

Nolt, John. 2015. "Casualties as a Moral Measure of Climate Change." *Climatic Change* 130, no. 3: 347–358.

Norton, Bryan. 1987. *Why Preserve Natural Variety?* Princeton, NJ: Princeton University Press.

Pimentel, David. 2006. "Soil Erosion: A Food and Environmental Threat." *Environment, Development and Sustainability* 8, no. 1: 119–137.

Pimentel, David, Michele Whitecraft, Zachary R. Scott, Leixin Zhao, Patricia Satkiewicz, Timothy J. Scott, Jennifer Phillips, Daniel Szimak, Gurpreet Singh, Daniela O. Gonzalez, and Tun Lin Moe. 2010. "Will Limited Land, Water, and Energy Control Human Population Numbers in the Future?" *Human Ecology* 38, no. 5: 599–611.

Pimm, S., C. Jenkins, R. Abell, T. Brooks, J. Gittleman, L. Joppa, P. Raven, C. Roberts, and J. Sexton. 2014. "The Biodiversity of Species and Their Rates of Extinctions, Distribution, and Protection." *Science* 344, no. 6187: 1246752. DOI: 10.1126/science.1246752.

Ripple, William, Christopher Wolf, Thomas Newsome, Mauro Galetti, Mohammed Alamgir, Eileen Crist, Mahmoud Mahmoud, and William Laurance. 2017. "World Scientists' Warning to Humanity: A Second Notice." *BioScience* 67, no. 12: 1026–1028.

Rolston, Holmes, III. 1988. *Environmental Ethics: Duties to and Values in the Natural World*. Philadelphia, PA: Temple University Press.

Ryerson, William. 2010. "Population: The Multiplier of Everything Else." In *The Post Carbon Reader: Managing the 21st Century's Sustainability Crises*, eds. Richard Heinberg and Daniel Lerch, 153–174. Healdsburg, CA: Watershed Media.

Sarkar, Sahotra. 2005. *Biodiversity and Environmental Philosophy*. Cambridge: Cambridge University Press.

Science and Security Board of the Bulletin of the Atomic Scientists. 2020. "Closer than ever: It is 100 seconds to midnight." *Bulletin of the Atomic Scientists*. https://thebulletin.org/doomsday-clock/current-time/. Accessed February 18, 2020.

Selby, Jan, Omar Dahi, Christiane Frohlich, and Mike Hulme. 2017. "Climate Change and the Syrian Civil War Revisited." *Political Geography* 60: 232–244.

Shragg, Karen. 2015. *Move Upstream: A Call to Solve Overpopulation*. Minneapolis–St. Paul, MN: Freethought House.

Siebert, S., J. Burke, J. M. Faures, K. Frenken, J. Hoogeveen, P. Döll, and F. T. Portman. 2010. "Groudwater for Use in Irrigation – a Global Inventory." *Hydrology and Earth System Sciences* 14, no. 10: 1863–1880.

Smail, J. Kennth. 1997. "Beyond Population Stabilization: The Case for Dramatically Reducing Global Human Numbers." *Politics and the Life Sciences* 16, no. 2: 183–192.

Springmann, Marco, Daniel Mason-D'Croz, Sherman Robinson, Tara Garnett, H. Charles J. Godfray, Douglas Gollin, Mike Rayner, Paola Vallon, Peter Scarborough. 2016. "Global and Regional Health Effects of Future Food Production Under Climate Change: A Modelling Study." *The Lancet* 387, no. 10031: 1937–1946.

Thomas, Chris, Alison Cameron, Rhys Green, Michel Bakkenes, Linda Beaumont, Yvonne Collingham, Barend Erasmus, Marinez Ferreira de Sequeira, Alan Gainger, Lee Hannah, Lesley Hughes, Brian Huntley, Albert van Jaarsveld, Guy Midgley, Lera Miles, Miguel Ortega-Huerta, A. Townsend Peterson, Oliver Phillips, and Stephen Williams. 2004. "Extinction Risk from Climate Change." *Nature* 427, no. 6970: 145–148.

UN Department of Economic and Social Affairs, Population Division. 2019. *World Population Prospects 2019: Volume I: Comprehensive Tables*. https://population.un.org/wpp/Publications/Files/WPP2019_Volume-I_Comprehensive-Tables.pdf. Accessed December 5, 2019.

UN Food and Agricultural Organization. 2006. *World Agriculture towards 2030/2050*. Rome: UN Food and Agricultural Organization. www.fao.org/fileadmin/user_upload/esag/docs/Interim_report_AT2050web.pdf. Accessed December 5, 2019.

UN Food and Agricultural Organization. 2015. *The State of Food Insecurity in the World*. Rome: UN Food and Agricultural Organization. www.fao.org/3/a-i4646e.pdf. Accessed December 6, 2019.

UN Food and Agriculture Organization. 2016. *The State of the World Fisheries and Aquaculture*. Rome: UN Food and Agricultural Organization. www.fao.org/3/a-i5555e.pdf. Accessed December 5, 2019.

Watson, Paul. 2012. "The Laws of Ecology and Human Population Growth." In *Life on the Brink: Philosophers Confront Population*, eds. Phil Cafaro and Eileen Crist, 130–137. Athens, GA: University of Georgia Press.

Weisman, Alan. 2013. *Countdown: Our Last, Best Hope for a Future on Earth?* New York: Little, Brown, and Company.

Wilson, Edward O. 1992. *The Diversity of Life*. Cambridge, MA: Belknap Press.

Wilson, Edward O. 2016. *Half-Earth: Our Planet's Fight for Life*. New York: Liveright.

World Health Organization. 2005. *Climate and health: Fact sheet, July 2005*. www.who.int/globalchange/news/fsclimandhealth/en/index.html. Accessed December 6, 2019.

World Health Organization. 2009. *Global Health Risks: Mortality and burden of disease attributable to selected major risks*. Geneva, Switzerland: WHO Press. www.who.int/healthinfo/global_burden_disease/GlobalHealthRisks_report_full.pdf. Accessed December 6, 2019.

World Health Organization. 2014. *Quantitative risk assessment of the effects of climate change on selected causes of death, 2030s and 2050s*. Geneva: WHO Press. http://apps.who.int/iris/bitstream/10665/134014/1/9789241507691_eng.pdf. Accessed December 6, 2019.

Worldometers. 2019. "World Population." www.worldometers.info/world-population/. Accessed December 5, 2019.

Zeebe, Richard. 2013. "Time-dependent Climate Sensitivity and the Legacy of Anthropogenic Greenhouse Gas Emissions." *Proceedings of the National Academy of Sciences of the United States of America* 110, no. 34: 13739–13744.

Zwarthoed, Danielle. 2016. "Should Future Generations be Content with Plastic Trees and Singing Electronic Birds?" *Journal of Agricultural and Environmental Ethics* 29: 219–236.

Part II

Intergenerational ethics, population policy, and personal procreative obligations

3 Intergenerational equity and long-term environmental impacts

The facts about our current environmental circumstances paint a dreary picture, but that need not lead to panic or despair. The point in understanding our circumstances is so that we can determine how to respond. Hundreds of millions of people this century may suffer severe harm from our ongoing environmental impacts, and billions could be threatened in the centuries that follow. Our large numbers are not the only source of these problems, but the large size of the human population makes all of these problems more difficult to tackle. In this part of the book, I present my primary ethical argument for the claim that we should be pursuing near-term population stabilization and long-term population reduction. I will then consider the implications of this argument with respect to both social policies and individual moral obligations.

My central argument, which I will hereafter call the *Population Reduction Argument* (PRA), is a series of interconnected smaller arguments. While these smaller arguments will all be defended in their own sections, I present the entirety of PRA here at the outset:

1 We morally ought to avoid causing massive and unnecessary harm to presently existing people.
2 Our moral duties of non-harm are just as stringent toward future people as they are toward present people.
3 Thus, we morally ought to avoid causing massive unnecessary harm to future people. [1, 2]
4 If we do not dramatically reduce our current levels of environmental degradation, then we will cause massive and unnecessary harm to future people.
5 Therefore, we morally ought to dramatically reduce our current levels of environmental degradation. [3, 4]
6 The anthropogenic environmental degradation caused by a human population is the product of the population size and the average rate of environmental degradation per person.
7 Thus, we morally ought to reduce our population size, reduce the average rate of environmental degradation per person, or reduce both

our population size and the average rate of environmental degradation per person. [5, 6]
8 There is no morally permissible way to reduce population size enough to adequately respond to our environmental problems if the average rate of environmental degradation per person remains unchanged.
9 There is no feasible way to reduce the rate of environmental degradation per person enough to adequately respond to our environmental problems if our population size remains at its current size or continues to grow.
10 Therefore, we morally ought to both reduce our rates of environmental degradation per person and reduce our current population size. [7, 8, 9]
11 We morally ought to reduce our current population size. [10]

In philosophical terms, we can say that PRA is logically valid. That means that if the premises of the argument are true, then its conclusion must be true. The premises are claims (1), (2), (4), (6), (8), and (9); all the other claims are conclusions that follow from these statements. Thus, if the argument has a flaw, it must be that one of those six premises is false.

In the remainder of this chapter, I defend the first half of PRA, which ends with claim (5). I defend the remainder of the argument in Chapter 4.

The basic duty not to harm

One of the cornerstones of morality is a commitment to avoid causing harm to other people. As with all basic moral principles, there are exceptions, such as punishing wrongdoers or violently opposing grave injustice. But the general rule remains: when we make a choice to harm someone – to impose on them pain, suffering, or death – we are expected to have a good justification for doing so. If I assault someone and my only reason is that I do not like the other person's appearance, I am not only blameworthy for my actions but may also face serious legal consequences. To justify my action, I would need a much stronger reason for harming the person. Perhaps if they were threatening my life or safety, my assault could be justified.

We can, of course, come up with plenty of other circumstances where imposing harm is morally acceptable, but in considering those cases, we are always starting from the assumption that harming someone is wrong. We only drift from this judgment when the party who imposes the harm has a compelling moral justification for their actions. For severe harms, the justification for causing them needs to be proportionally strong. Killing someone, for instance, will only be justifiable in the most extreme of circumstances, such as when your life or the lives of innocent others are at risk. For minor harms, weaker justifications may suffice. Grounding your child may make them unhappy and thereby harm them, but it can be an appropriate punishment for even small transgressions, such as telling a lie or saying something disrespectful, because the harm imposed is mild and brief.

With these basic observations about harm in mind, we can now turn to the starting point for PRA:

1   We morally ought to avoid causing massive and unnecessary harm to presently existing people.

I define harm as "massive" when it causes death and/or various forms of significant suffering (e.g., starvation, life-threatening illness, psychological distress, severe pain) to large numbers of people. I define harm as "unnecessary" when it is not imposed to protect people's lives, welfare, or other morally significant interests. Thus, claim (1) states a moral obligation not to kill, maim, injure, imprison, or otherwise severely harm other people without a very strong justification for doing so. This obligation is fundamental to any plausible moral theory or code of morality. To deny it would require being a skeptic about there being any moral truths at all.[1]

Equity of non-harm

With my support for claim (1) established, we can turn to the second premise in PRA, which I will refer to as *Equity of Non-Harm*:

2   Our moral duties of non-harm are just as stringent toward future people as they are toward present people.

Whereas claim (1) is a widely taught moral precept, *Equity of Non-Harm* is not. In fact, some may recoil from the principle immediately. They might reason that present people exist while future people do not and that this represents a tremendous moral difference between these two groups of people. Present people are already here, and so there is no doubt that they matter to our moral decision-making. But future people may never exist, and so they *might not* matter to our moral decision-making. This could provide a reason for regarding our duties of non-harm as being stronger toward present people than future people.

One cause of this reaction could be confusion about what I mean when use the phrase "future people." I define a *future person* to be someone who *will* exist at some later point in time but does not exist now.[2] Future people should not be conflated with *merely possible people*. A merely possible person is someone who could exist but never actually comes into existence. Merely possible people should not play a crucial role in our moral decision-making because they have no interests and never will have any interests: there is nothing we can do to affect them for good or for ill. But future people *do* have interests. Minimally, they have interests in being provided the means to live a decent human life. For this reason, we can undertake actions that promote or threaten their interests. Our way of discussing future people often tacitly affirms this idea. Consider a married couple contemplating how they could ensure their not-yet-conceived children can obtain a good education. In much the same way, we can assess how our actions could promote or threaten the interests of people who will inhabit the Earth in 50 or 100 years. Moreover, we know enough about biology, psychology, economics, and related fields to reliably anticipate some of their interests: clean air, sanitized drinking water, reliable food sources, a habitable environment, protection from physical harm, educational opportunities, a job with a livable wage, affordable medical care, and so on. The list is familiar because these are the same things that present people desire. Thus, if there is a relevant moral difference between present and future people, it is *not* that future people lack interests or that we don't know what their interests are.

An alternative objection to *Equity of Non-Harm* originates from contract theories of ethics. According to these theories, moral principles are rooted in "mutually agreeable reciprocity of cooperation between individuals" (Darwall 2002, p. 1). On some understandings of this view, we cannot have obligations to future people because we cannot interact with them in the ways needed to form reciprocal agreements. If this thought is correct, then we cannot have any moral obligations – including those of non-harm – to people who do not yet exist.

The challenges with extending contract theories to future people and intergenerational moral issues are well-documented (e.g., Ashford and Mulgan 2018; Fineron-Burns 2017; Gardiner 2009; Frick 2015). If contract theories are ultimately limited to explaining moral obligations between contemporaries, then this fact would only reveal that such theories are unsuitable for addressing intergenerational moral questions. If we had no obligations at all to those who do not yet exist, then we would be unable to explain some of our most fundamental beliefs about the morality of procreation. As one illustration, if we have no moral duties to future people, then parents would not be required to refrain from conceiving a child whose life will most likely be dreadful (e.g., due to a debilitating genetic disease), even if it were easy to prevent the child's conception. Moreover, no impact on distant future people – no matter how severe – could be considered morally wrong. This result is unacceptable because there are clear cases where it is wrong to cause future people to suffer or die. Let's consider a thought experiment to illustrate this point.

Imagine a despicable terrorist who has adopted the alias Coda. One day, Coda plants a time bomb on the ground floor of an elementary school. It explodes hours later and kills ten students. Undeniably, this act is wrong: in fact, it is among the most morally wretched things a person could do. Now imagine the same scenario but with one change: Coda rigs the timer with a long-lasting battery and schedules the detonation for 15 years later. The bomb is well-hidden and goes undiscovered. It explodes 15 years later and kills ten students. As it happens, all of the students who died were born years after Coda planted the bomb. Does it make any moral difference that the casualties in the second case were future people at the time the bomb was planted? If we assume that Coda had good reason to think the school would still exist in 15 years, then this fact seems to make no difference: his action in this case is just as wrong as the case in which the bomb detonates shortly after being planted. The deaths of ten children 15 years into the future are just as bad as the deaths of ten children in the present.

Another consideration that points to the moral insignificance of temporal distance is the moral insignificance of spatial distance. Where a person is located does not affect their moral value, rights, interests, or the degree to which we should respect their rights and welfare. Their geographical location is an arbitrary fact (in large part determined by factors beyond their control) that says nothing about their character, rational capacities, or moral agency. Arbitrary considerations like this do not provide legitimate justifications for regarding other people as morally less important than ourselves.[3]

At this juncture, one might object that geographical distance actually does matter to what we owe other people because it affects our relationships with them. Typically, we hold that our special relationships with our families and members of our local communities create obligations that we do not have to distant strangers. In the same manner, our temporal distance from future people seems to restrict our relationship with them since we are unable to have any kind of reciprocal interaction. Perhaps this difference could explain why our duties not to harm future people are not as strong as our duties not to harm present people.

While it is true that we generally have stronger duties to those who are geographically nearer to us, this fact does not result from geographical distance as such. What matters in such cases are the *relationships* that we form with others. After all, we can have special duties to friends or family members who are geographically distant from us, and advances in transportation and communication have made these relationships relatively common. The real question is this: is the relationship that we have with future people sufficiently different from our relationship with present people to justify our having less stringent duties of non-harm? The answer is no.

It is true that we have special duties to the small portion of presently existing people with whom we form strong interpersonal relationships, but we also have moral duties that extend to *all* people, including those we have never met and never will meet. These include duties not to steal their property, physically assault them, jeopardize their welfare, or otherwise cause them harm. These obligations cannot plausibly be grounded in any relationship we form with all these people. The notion that we form a relationship with all presently existing people – even a very loose one – is dubious. We form morally significant relationships with only a very small portion of those who presently exist.

To the extent that special relationships affect our moral duties, they are usually taken to only affect our duties of assistance. I may well have duties to assist my friends and family (e.g., doing favors for them, helping them in emergencies, providing financial assistance to my children) that I do not have to strangers. But our duties not to harm or wrong others are usually taken to apply to everyone – strangers and close acquaintances alike. Just as it is wrong for me to assault one of my close friends (except in unusual circumstances), it is wrong for me to assault a stranger. I should avoid harming other people regardless of my relationship with them. *Equity of Non-Harm*, as the name implies, only applies to duties not to harm others, so its stringency is not affected by whether or not we have personal relationships with others.

A further consideration in favor of *Equity of Non-Harm* originates from the general trend in ethics to ground moral status in the capacities that a person or animal possesses. Many theorists in animal ethics have approached what we owe to nonhuman animals as dependent upon the animals' morally relevant capacities (DeGrazia 1996; Midgley 1983; Nussbaum 2006, ch. 6; Regan 1983; Sapontzis 1987; Singer 2002). These authors appeal to the capacities of animals to establish moral obligations to them. First, they identify what features serve to ground human beings' moral status, and then they demonstrate that animals have some of the same morally relevant capacities. Philosophers do not all agree about what capacities are important, but this list often includes the capacity to feel pleasure and pain, to exercise autonomy, to make rational decisions, to understand oneself as an entity that continues to exist through time, to engage in linguistic communication, and to establish meaningful relationships with others.

According to these capacity-oriented accounts of moral status, if it is wrong to harm people who have certain morally relevant capacities, then it is also wrong to harm animals that possess the same morally relevant capacities. We can apply the same reasoning to future people. If it is wrong to harm presently existing people and future people have the same morally relevant capacities as presently existing people, then it is likewise wrong to harm future people. Will future people have the same morally relevant capacities that we do? If we are speculating about the characteristics of human beings who might exist in a million years, then it might be reasonable to think that such people would be significantly different from us. But on the shorter time scales in which we are considering our intergenerational moral duties, there can be no doubt that future people will be like us in all the morally relevant ways. They will have the same general psychological and biological characteristics that we possess. We have no reason to believe, for instance, that those living in 2100 will have lost the ability to reason or will have become unable to experience pleasure and pain. Since future people – at least those who will exist during the next several centuries – will have the same morally relevant capacities as present people, our duties not to harm future people are just as strong as those we have not to harm presently existing people.

These preceding considerations generate strong support for *Equity of Non-Harm*, but some may be reluctant to accept my reasoning. Two major objections need to be addressed. The first stems from a concern about how our actions now affect the identities of those who exist in the future. The other arises from the widespread use of the social discount rate in economic decision-making. I will address each in turn.

The non-identity problem

Sometimes, our actions in the present have an effect on the identities of the people who will exist in the future. This observation forms the core of a philosophical conundrum known as the Non-Identity Problem (NIP). Its most well-known presentation originates with Derek Parfit (1987, pp. 351–379), but Gregory Kavka (1982), Robert Adams (1979), and Thomas Schwartz (1978) also discovered and articulated the problem in their work. Since NIP has generated a robust literature and poses a direct challenge to the second premise of PRA, it warrants an explanation and response.

Parfit (1987) starts his presentation of NIP by defending the following principle, which he calls the Time-Dependence Claim: "If any particular person had not been conceived within a month of the time when he was in fact conceived, he would in fact never have existed" (p. 352). Because our genes form an essential part of our identity, the timing of our birth matters a great deal to who we are. Had my parents conceived a child a month later than I was conceived, *I* would not have been born. My parents would have still had a child, but that child would have had a different genetic makeup. As a result, this child would not have been *me* but a completely different person.

The Time-Dependence Claim matters in the context of intergenerational ethics because of an assumption some make about the notion of harm. Many adopt a counterfactual comparative account of harm (CCH). According to CCH, a person is harmed by an action only if that action makes the person worse off than they otherwise would have been. So, to determine whether an action caused harm to someone, we must consider what would have happened if the action were not performed and compare that outcome to what actually happened. If I break your arm, then you could claim that I harmed you because you would have avoided serious pain and discomfort if I had done nothing. My action, in other words, made you worse off than you would otherwise be.

If we endorse both CCH and the Time-Dependence Claim, then future people sometimes appear incapable of being harmed by our actions. Consider a case of long-term environmental policy, which I will label *Depletion*:

> As a community, we must choose whether to deplete or conserve certain kinds of resources. If we choose Depletion the quality of life over the next two centuries would be slightly higher than it would have been if we had chosen Conservation. But it would later, for many centuries, be much lower than it would have been if we had chosen Conservation.... Suppose that we choose Depletion. Is our choice worse for anyone?
>
> (Parfit 1987, pp. 362–363)

40 *Ethics, policy, and obligations*

When most of us contemplate *Depletion*, we have a strong conviction that there is something seriously wrong with depleting resources and making the world so much worse for future people. But if we choose to conserve resources, then the people who exist will be completely different than the ones who exist in the scenario where we deplete resources. As a result, the people who live in a depleted world cannot claim that our choice to deplete resources made them worse off than they otherwise would have been. Thus, so long as they have lives that are worth living (i.e., welfare high enough that it is better to live than not exist), our actions do not appear to have harmed them. This is the core of NIP: how do we explain our conviction that depleting resources is wrong when doing so does not make anyone worse off than they otherwise would have been?

NIP is particularly salient in this context because *Depletion* has some parallels to policies we might pursue to mitigate climate change or reduce population. Suppose that we were considering whether to implement an international policy that only allows couples to have a maximum of two children. Under full compliance to a global two-child policy, the world's population would eventually stabilize and then gradually reduce, but this outcome would take some time to occur. Suppose that stabilization will not occur for 100 years. Under this policy, many people will have children at different times or refrain from having children that they otherwise would have had, and some people will meet and start families together who would otherwise never have met. After several generations, these outcomes could result in a completely different set of people existing in 100 years than would have existed under a less restrictive procreative policy. But this means that the people who would have existed under that less restrictive policy, given that their lives would have still been worth living, would not have been harmed by that choice. In fact, they *needed* us to choose that policy in order to exist at all. Since they would not have been made worse off, it appears that permitting a more liberal procreative policy – even one that lowered overall welfare by exacerbating overpopulation – would not have harmed any future people.

NIP is an intriguing challenge to applying principles of non-harm to intergenerational cases, but it has some practical limitations that we should acknowledge. First, most of the environmental problems that concern us here are already underway. They are having impacts on people who already exist and some who will exist regardless of what policies we adopt in the near future. Since NIP is only applicable to effects that are in the distant future, it has limited import to our current circumstances.[4] Second, an assumption of NIP is that the people in the future will still have lives worth living – that their lives will not be so bad that nonexistence would be preferable. If a future person has a life so plagued by suffering that nonexistence would be preferable, then they might still be able to claim that our actions made them worse off even if those actions were necessary for them to exist. They could reason that if they did not exist, their welfare would be equivalent to zero (or

Equity and long-term environmental impacts 41

nothing) and that their current circumstances are worse than nothing.[5] So by being brought into existence, they were made worse off. If the impacts of climate change, biodiversity loss, and other environmental impacts are severe enough, then some future people may have lives that are not worth living. Under those conditions, they could claim that our actions harmed them even if their identities were dependent on our acting as we did and even if CCH were the correct account of harm.

These caveats notwithstanding, there is also a more direct response to NIP available: we can just reject CCH. What these identity altering cases suggest is that CCH is an incomplete concept of harm. It might well be a *sufficient* condition for harming someone: if you make a person worse off than they would have otherwise been, they have pretty straightforward grounds for claiming that you harmed them. But it does not appear to be a *necessary* condition for harming someone. Rather than constituting a genuine threat to intergenerational ethics, NIP just points to a shortcoming in CCH – one kind of scenario where a person can be harmed without being made worse off than they otherwise would have been.

This point becomes clearer if we reflect on the kind of conclusions we will have to endorse if we accept the implications of NIP. In one of the first presentations of NIP, Gregory Kavka (1982) presents the case of parents who conceive a child specifically for the purpose of selling the child into slavery to make a profit (p. 100). Suppose the child is conceived and then immediately taken by the slaveholder as a newborn. It is clear that the parents have done something wrong, but if they would not have conceived a child otherwise and if the child's life as a slave is better than nonexistence, then the reasoning that underlies NIP suggests that they did not do anything wrong. After all, the child has not been made worse off by this arrangement, and the other parties have all benefited. Similar reasoning will apply in cases where parents could avoid conceiving a child with a genetic impairment by conceiving a child later (perhaps after undergoing some form of medical treatment).[6] If they decide to conceive a child immediately, then NIP will entail that they did nothing wrong since that particular child would not have been born otherwise.

Accepting these conclusions would be a mistake. Instead, we should reject CCH and acknowledge that this conception of harm is inadequate for approaching issues in intergenerational ethics. That result should not surprise us because this understanding of harm is most applicable in everyday scenarios involving actions between presently existing people. It would be more surprising if CCH were equally useful in an intergenerational context when it was developed outside that context. We should not be reluctant to abandon it.

While I will now turn my attention to the second major objection to *Equity of Non-Harm*, I recognize that some philosophers will deem this dismissal of NIP too superficial. If you are among those readers, I invite you to read the far more elaborate response to NIP contained in the Appendix.[7]

The social discount rate

The second major objection to *Equity of Non-Harm* originates from the economic practice of discounting the future. Many contemporary policy-makers assume that the significance of future events decreases over time. A social discount rate is a percentage by which future outcomes, whether good or bad, decrease in value at a fixed percentage each year. The underlying rationale has its roots in economics. Imagine we can invest our money with a 5 percent return on our investment, and suppose we are given the choice between receiving $1000 now or $1000 one year in the future. Under these circumstances, it would be preferable to receive the money immediately. We could invest $953, and in one year, this money will total just over $1000 because of our 5 percent return on our investment. And in the meantime, we have $47 to spend.

In a purely economic context, the use of a social discount rate often makes sense, but it does not make sense to apply the discount rate to morally significant outcomes. Ten people dying in a year is not 5 percent less bad than ten other people dying right now. These outcomes are equally bad. The mere passage of time does not indicate that the moral significance of an event decreases. Usually, the purported justifications for a social discount rate reveal that the passage of time is not the real reason for discounting. For example, some justify the social discount rate on the basis of distant future events having a lower probability of occurring than events that will occur in the immediate future. Granting this claim about the relative probability of events, however, does not entail that future benefits and losses are *less valuable* than they are in the present; instead, what this would demonstrate is that it is harder to predict benefits and losses the further into the future they are projected to occur. These benefits and losses would still be just as morally significant in the future as they are in the present, if they were to occur. Harms that occur to future people are not less significant *just because* they take place in the future.

Rather than masking our concern about probabilities behind a social discount rate, we should just directly assess what the probabilities of future events occurring are. With regard to the environmental problems that we are considering, an appeal to low probability of occurrence would be a very poor reason for being unconcerned about future outcomes. Their probabilities are not low or uncertain: the harms are already observable, and we have no reason to think they will cease unless some aspects of our behavior change drastically.

Another attempt to justify a social discount rate relies on the claim that future people will (overall) be better off than we are and that harms they suffer will therefore matter less to them than they do to us. This argument does not fare much better than the appeal to the uncertainty of future events occurring. Some harms, such as death or severe suffering, are grave enough that they will probably matter just as much to future people as us. Moreover, given the rises and falls of civilizations over the course of human history, it is uncertain whether the future will be better than the past in the linear manner that would justify a uniform social discount rate.

There is, however, one defense of the social discount rate that is more sophisticated than these prior arguments – one that draws on considerations related to NIP. Duncan Purves (2016) argues that the social discount rate is justified because of the ways that our actions in the present affect the identities of those in the future. Importantly, although Purves often speaks as if CCH is correct, he does not rely on it being a true and comprehensive account of what actions qualify as harms. Instead, his central claim is that counterfactual-comparative harms are *morally worse* than (merely) non-counterfactual comparative harms (p. 218). To illustrate this point, Purves evaluates Hanser's (2008) event-based account of harm, according to which people are harmed when they are deprived of basic goods. Purves (2016) thinks that accounts like Hanser's – that is, accounts that do not invoke CCH – will be unable to explain the intuitive verdict about this case, which he calls *Burning Building*:

> George sees two of his neighbors trapped in a burning building. Jane, one of the people trapped in the building, has a fatal heart condition such that if she is not killed by the fire, she will die from her heart conditions moments later. Elroy, the other people [sic] trapped in the building, has no such heart condition. If he is not killed by the fire, he will enjoy many more good years of life.
>
> (p. 217)

According to Purves, if George has full information about each of these people, he ought to save Elroy, and the reason he ought to save Elroy is that Elroy would suffer a greater harm by dying in the fire than Jane would. Elroy would be deprived of many good years of life if he died in the fire whereas Jane would be deprived of only a short bit more life. But on Hanser's account, both Elroy and Jane are harmed to the same degree by dying in the fire because they would lose all their basic goods. That verdict is problematic, according to Purves (2016): "clearly the harm to Elroy is greater, and it would be morally worse" (p. 219).

We can (and should) grant Purves' claim that we should save Elroy in the Burning Building case, and this case illustrates well enough why CCH is "something we should care about" (Purves 2016, p. 217). If we do not save Elroy, then he will be deprived of life he otherwise could live whereas Jane is not similarly deprived of life if she is not saved. Thus, there are circumstances where CCH matters to our moral evaluations. Nonetheless, all this can be granted without endorsing Purves' conclusions about the moral significance of harms to future people. He does not provide sufficient reason for thinking that counter-factual comparative harms are always worse than harms he would classify as non-comparative.

Consider a couple, Tom and Sandy, living in the future. Both die in a tropical storm resulting from climate change. As it happens, Tom would have been born whether or not we enacted policies to significantly mitigate climate change, but Sandy would not have been born if we had done so. According to CCH, Tom is harmed by the storm and Sandy is not even though they are victims of the same event and both see their lives end as a result of it. Purves does not demonstrate that the harm to Sandy is less significant than the harm to Tom in this type of scenario. In *Burning Building*, saving Jane will yield no meaningful moral benefit – the outcome will be the same for her whether she is saved or not. But if actions are taken to mitigate climate change so that Tom is spared by the storm, there will also be a morally significant benefit to someone else. That someone will not be Sandy (because she will not exist), but suppose Tom marries Amanda instead. Amanda will not die in that tropical storm, and the outcome is morally better when Tom's wife – regardless of her specific identity – does not die prematurely in this disaster. Thus, preventing the non-comparative harm to Sandy is morally significant in a way that preventing the non-comparative harm to Jane in *Burning Building* is not. For this reason, we can agree with Purves assessment of *Burning Building* but deny his more general claim that counterfactual comparative harms are more significant than non-counterfactual comparative harms.

One reason the intuitive appeal of CCH disappears in intergenerational cases is that it is being applied with an unusually broad scope. CCH, as employed in non-identity cases, "automatically aggregates all the consequences of an action and determines on the basis of the resulting ensemble whether the action has caused harm" (Nolt 2013, p. 115). In other contexts, CCH is never employed this way. Imagine that Jerry is crossing the street, and I strike him with my car while driving recklessly. He breaks his leg in several places and spends several weeks in a local hospital while recovering. As it happens, however, he falls in love with one of the nurses there, someone he would not have met if he had not been admitted to the hospital. His love is ultimately reciprocated, and the two enjoy a lasting, loving relationship until they die. Did I harm Jerry by striking him that fateful day with my car? If we apply CCH in a very broad way, then it appears I did not, since my breaking his leg bestowed on him a benefit in the long term that was greater than the harm he initially suffered. But such an analysis is not consistent with how such a case would typically be judged. Rather, it seems that my action caused two distinct effects – the harm of Jerry breaking his leg and the benefit of him finding true love.

The peculiarity of applying CCH in this broad, effect-aggregating manner becomes apparent when we consider its implications. If we were to always apply CCH this way, then we would be fraught with uncertainty about whether many actions were really harmful. Knowing whether something was really a harm would require us to know *all* the action's long-term effects so that we could compare the harms and benefits accrued by the action. Since we virtually never have such knowledge, the concept of harm would become useless in our moral reasoning. Thus, when we apply CCH, we almost always apply it more narrowly. My striking Jerry with my car makes him worse off in the sense that he now has a broken leg, not in the sense that his *entire life* is now worse on the whole as a result of this event.

The same analysis can be applied to identity-affecting cases. Let's return to Tom and Sandy's deaths in the tropical storm. Our failure to mitigate climate change harms Sandy by causing her death, and it also benefits Sandy by providing the conditions necessary for her existence. These are distinct effects and should not be aggregated together. Just as Jerry being benefited does not erase the harm he suffered, Sandy being benefited does not erase the harm she suffers. Ultimately, we can understand the harm that Sandy suffers in either a non-comparative way or in a narrowly comparative way. On either account, the harm she suffers is morally significant and just as morally significant as counterfactual-comparative harm that Tom suffers.

Overall, the prospects for justifying a social discount rate are not promising (Kelleher 2012; Nolt 2015, pp. 96–201; Parfit 1987, pp. 480–486). We should regard events that will occur far in the future as having the same moral weight as those that will occur soon. This completes my defense of PRA's second premise. From the argument's first two premises, we can derive the following claim:

3   We morally ought to avoid causing massive unnecessary harm to future people.

The final argumentative step of this chapter is connecting this abstract moral claim to our real-world circumstances.

The link to environmental degradation

In Chapter 2, I examined some of the ongoing environmental problems that threaten the welfare of both present and future people. Among others, these include global climate change, biodiversity loss, regional freshwater shortages, and regional food shortages. Thus, the previous chapter provides ample support for the next step in PRA:

4   If we do not dramatically reduce our current levels of environmental degradation, then we will cause massive and unnecessary harm to future people.

Since these environmental impacts could cause severe suffering and death to hundreds of millions (if not billions) of future people, it should be clear that these harms will be "massive" on any plausible meaning of the term.

These harms are unnecessary because we do not need to engage in the actions that cause them. Someone might try to argue that our high rates of

46 *Ethics, policy, and obligations*

consumption are necessary to maintain our welfare, but such a claim is dubious. First, for those of us in the developed world, many of our environmentally destructive behaviors are not essential to improving our well-being (Andreou 2010; Gambrel and Cafaro 2010; Gardiner 2012, pp. 244–245). Often, these behaviors stem from adherence to perceived social expectations or an unwillingness to break old habits when more eco-friendly behaviors could promote a person's health and well-being just as well as their current ones. Second, even assuming that many of the consumption-driven activities do increase people's welfare, it does not follow that the activities are *necessary* in any morally meaningful sense. The fact that stealing someone's property would improve my welfare does not entail that it would be morally justified. Some environmentally destructive activities really are necessary for people to survive in their current circumstances, but many of the activities under consideration here do not remotely approach that level of need. We do not need our homes kept at a stable temperature of 72 degrees year-round, and we do not need to purchase large, fuel-inefficient vehicles just because their appearance is appealing. We could refrain from these activities with only marginal costs to our well-being. Thus, the harms resulting from these and similar activities are unnecessary.

Linking claims (3) and (4) together brings us to the end of this chapter. We have arrived at the following conclusion:

5   We morally ought to dramatically reduce our current levels of environmental degradation.

The main thought expressed in (5) is one that a lot of environmental scientists, activists, and ethicists share. Yet, even though many people share this conviction, only a small portion of them have advocated for reducing our levels of environmental degradation by trying to reduce our population size. In the next chapter, I will proceed from claim (5) to the conclusion that we need to pursue population reduction as one means of responding to our current environmental problems.[8]

Notes

1 Here and throughout the book, I am assuming that moral truths exist and that it is possible to make progress in answering ethical questions. A robust response to moral skeptics is beyond my means here, and since they doubt that any moral claims are true, I imagine they will have little interest in the project of applying moral principles to issues in environmental ethics.
2 I borrow this usage from Nolt (2016).
3 For further reasons to regard temporal distance as morally insignificant in the context of intergenerational ethics, see Caney (2014, pp. 323–325) and Nolt (2019).
4 This is the main reason that David Boonin (2014), one of the biggest defenders of NIP, acknowledges that non-identity considerations may not have significant practical implications even if we endorse the main claims that give rise to NIP (p. 216).

5 Whether one can plausibly assign a welfare value of zero to nonexistence is controversial. For some defenses of this position, see Feldman (1991), Holtug (2001), and Roberts (2003).
6 David Boonin (2014) models his own treatment of NIP on a case of this type.
7 I thank an anonymous reviewer for the suggestion to gather this material into a standalone appendix.
8 Significant portions of this chapter are derived from chapter 5 of my doctoral dissertation. See Hedberg (2017).

References

Adams, Robert. 1979. "Existence, Self-Interest, and the Problem of Evil." *Nous* 13, no. 1: 53–65.

Andreou, Chrisoula. 2010. "A Shallow Route to Environmentally Friendly Happiness: Why Evidence That We Are Shallow Materialists Need not be Bad News for the Environment(alist)." *Ethics, Place and Environment* 13, no. 1: 1–10.

Ashford, Elizabeth, and Tim Mulgan. 2018. "Contractualism." *Stanford Encyclopedia of Philosophy*. https://plato.stanford.edu/entries/contractualism/. Accessed September 28, 2019.

Boonin, David. 2014. *The Non-Identity Problem and the Ethics of Future People*. Oxford: Oxford University Press.

Caney, Simon. 2014. "Climate Change, Intergenerational Equity and the Social Discount Rate." *Politics, Philosophy & Economics* 13, no. 4: 320–342.

Darwall, Stephen. 2002. "Introduction." In *Contractarianism/Contractualism*, edited by Stephen Darwall, 1–8. Malden, MA: Blackwell.

DeGrazia, David. 1996. *Taking Animals Seriously: Mental Life and Moral Status*. Cambridge: Cambridge University Press.

Feldman, Fred. 1991. "Some Puzzles about the Evil of Death." *The Philosophical Review* 100, no. 2: 205–227.

Frick, Johann. 2015. "Contractualism and Social Risk." *Philosophy & Public Affairs* 43, no. 3: 175–223.

Finneron-Burns, Elizabeth. 2017. "What's Wrong with Human Extinction?" *Canadian Journal of Philosophy* 47, nos. 2–3: 327–343.

Gambrel, Joshua Colt, and Philip Cafaro. 2010. "The Virtue of Simplicity." *Journal of Agricultural and Environmental Ethics* 23, nos. 1–2: 85–108.

Gardiner, Stephen. 2009. "A Contract on Future Generations?" In *Intergenerational Justice*, edited by Axel Gosseries and Lukas Meyer, 77–118 Oxford: Oxford University Press.

Gardiner, Stephen. 2012. "Are We the Scum of the Earth?" In *Ethical Adaptation to Climate Change: Human Virtues of the Future*, edited by Allen Thompson and Jeremy Bendik-Keymer, pp. 241–259. Cambridge, MA: MIT Press.

Hanser, Matthew. 2008. "The Metaphysics of Harm." *Philosophy and Phenomenological Research* 77, no. 2: 421–450.

Hedberg, Trevor. 2017. "Population, Consumption, and Procreation: Ethical Implications for Humanity's Future." Ph.D. dissertation, Department of Philosophy, University of Tennessee.

Holtug, Nils. 2001. "On the Value of Coming into Existence." *The Journal of Ethics* 5, no. 4: 361–384.

Kavka, Gregory. 1982. "The Paradox of Future Individuals." *Philosophy & Public Affairs* 11, no. 2: 93–112.
Kelleher, J. Paul. 2012. "Energy Policy and the Social Discount Rate." *Ethics, Policy & Environment* 15, no. 1: 45–50.
Midgley, Mary. 1983. *Animals and Why They Matter*. Athens, GA: University of Georgia Press.
Nolt, John. 2013. "Replies to Critics of 'How Harmful Are the Average American's Greenhouse Gas Emissions?'" *Ethics, Policy & Environment* 16, no. 1: 111–119.
Nolt, John. 2015. *Environmental Ethics for the Long Term: An Introduction*. New York: Routledge.
Nolt, John. 2016. "Future Generations in Environmental Ethics." In *The Oxford Handbook of Environmental Ethics*, eds. Stephen Gardiner and Allen Thompson, 344–354. Oxford: Oxford University Press.
Nolt, John. 2019. "Domination across Space and Time: Smallpox, Relativity, and Climate Ethics." *Ethics, Policy & Environment* 22, no. 2: 172–183.
Nussbaum, Martha. 2006. *Frontiers of Justice: Disability, Nationality, and Species Membership*. Cambridge, MA: Harvard University Press.
Parfit, Derek. 1987. *Reasons and Persons*. Oxford: Oxford University Press.
Purves, Duncan. 2016. "The Case for Discounting the Future." *Ethics, Policy & Environment* 19, no. 2: 213–230.
Regan, Tom. 1983. *The Case for Animal Rights*. Berkeley, CA: University of California Press.
Roberts, Melinda. 2003. "Can It Ever Have Been Better to Have Existed at All? Person-Based Consequentialism and a New Repugnant Conclusion." *Journal of Applied Philosophy* 20, no. 2: 159–185.
Sapontzis, Steve. 1987. *Morals, Reason, and Animals*. Philadelphia, PA: Temple University Press.
Schwartz, Thomas. 1978. "Obligations to Posterity." In *Obligations to Future Generations*, edited by R. I. Sikora and Brian Barry, pp. 3–14. Philadelphia, PA: Temple University Press.
Singer, Peter. 2002. *Animal Liberation*. New York: HarperCollins Publishers, Inc.

4 The moral duty to halt population growth

In the previous chapter, I introduced Population Reduction Argument (PRA). The PRA is a lengthy string of claims that leads to the ultimate conclusion that we have a moral duty to reduce our population size. At the end of Chapter 3, I had supported 3 of PRA's 6 premises and ended with an endorsement of the following claim:

5   We morally ought to dramatically reduce our current levels of environmental degradation.

Here are the remaining steps in PRA:

6   The anthropogenic environmental degradation caused by a human population is the product of the population size and the average rate of environmental degradation per person.
7   Thus, we morally ought to reduce our population size, reduce the average rate of environmental degradation per person, or reduce both our population size and the average rate of environmental degradation per person. [5, 6]
8   There is no morally permissible way to reduce population size enough to adequately respond to our environmental problems if the average rate of environmental degradation per person remains unchanged.
9   There is no feasible way to reduce the rate of environmental degradation per person enough to adequately respond to our environmental problems if our population size remains at its current size or continues to grow.
10 Therefore, we morally ought to both reduce our rates of environmental degradation per person and reduce our current population size. [7, 8, 9]
11 We morally ought to reduce our current population size. [10]

Claims (6), (8), and (9) represent the remaining premises in the argument, so this chapter will focus entirely on defending those claims. I start with claim (6).

50 *Ethics, policy, and obligations*

Isolating the population and consumption variables

Claim (6) in PRA isolates the two main variables that combine to produce environmental degradation. The first is the population size, and the second is the average rate of environmentally harmful consumption within the population. We can write this as a simple equation:

> anthropogenic environmental degradation = population × average rate of environmental degradation per person

This second variable is not meant to be an approximation of every person's actual individual ecological footprint. Some environmental degradation results from how societies are structured and how collective entities (e.g., corporations) operate, and these impacts are not easily traceable back to the actions of particular individuals. Rather, my aim with this second variable is to isolate a broad measure of a society's rate of environmentally destructive behavior relative to its population size – something that could be considered a placeholder for the rate that a society is engaging in overconsumption.

This formulation varies slightly from the IPAT equation, which is one of the standard ways of understanding environmentally destructive impact.[1] According to the IPAT equation, environmental impact (I) is the product of population (P), affluence (A), and technology (T). My equation effectively combines affluence and technology into a single variable. This makes my equation simpler, but I have a different reason for favoring mine over IPAT.

The central problem with IPAT is that affluence and technology make variable contributions to environmental degradation. Improvements in technology, for instance, can result in more efficient use of resources, which reduces environmental impact. But often these improvements result in greater aggregate consumption because the economic demand for the resource increases, which increases environmental impact.[2] So whether technology boosts or lowers environmental impact will depend on the context. The same is true for affluence. While affluence more frequently results in increased environmental impact, it can also translate to greater financial resources to invest in renewable energy and green technologies. Thus, whether affluence boosts or lowers environmental impact will depend on the context. For these reasons, our analysis will be more fruitful if we focus on differentiating population size from the average rate of degradation per person rather than trying to parse out the specifics of affluence and technology in their particular contexts.

Claim (6) is not in itself a particularly interesting part of PRA, since it is little more than a mathematical equation. Its implications are what matter. Claims (5) and (6) lead us to the following conclusion:

7   We morally ought to reduce our population size, reduce the average rate of environmental degradation per person, or reduce both our population size and the average rate of environmental degradation per person.

Since the product of population size and average rate of environmental degradation produce the total amount of environmental degradation, it follows that we have three options for reducing environmental degradation: (1) only reduce population size, (2) only reduce the average rate of environmental degradation per person, and (3) reduce both population size and average rate of environmental degradation per person simultaneously. I will ultimately argue that the only viable option among these choices is (3).

Let's consider the possibility of only reducing population size. Recall from Chapter 2 that the global population is still growing and that we will most likely have close to 11 billion people on Earth in 2100 (UN Department of Economic and Social Affairs 2019). Remember that according to this option, we are *not* going to reduce the average rate of environmental degradation per person at all. Given our current population size and its projected growth rate, the only way to reduce the population sufficiently to lower environmental degradation with this stipulation would be to annihilate a large portion of the human population – literally billions of people. Perhaps that could be achieved in a worldwide nuclear war or if we developed a highly contagious lethal pathogen that was distributed at strategic locations around the world, but such measures would be utterly unconscionable. The only morally permissible way to reduce population is to lower fertility rates: eliminating people who already exist cannot be part of a morally satisfactory solution. But if the rates of environmental degradation per person do not change, then the rapid, large-scale reduction in global population that we would need to reduce environmental degradation is not obtainable – lowering fertility will not yield fast enough results for that to happen. These straightforward observations lead us to claim (8):

8   There is no morally permissible way to reduce population size enough to adequately respond to our environmental problems if the average rate of environmental degradation per person remains unchanged.

Anyone who understands the basic ethical and demographic aspects of our situation should accept this claim.

We are now left with two options: either we focus solely on reducing the average rate of environmental degradation per person or we try to both reduce the average rate of environmental degradation per person and reduce our population size. Now we must consider whether changing environmentally destructive behaviors could provide a satisfactory response to the problem even if we do not take steps to reduce our population size.

The need to reduce both of both population and consumption

The proposal to curtail environmentally destructive behavior, especially in the developed nations where per capita rates of consumption are high, has

52 Ethics, policy, and obligations

been the most popular response to our current environmental predicament. As discussed in the opening chapter, explicit discussions of population over the last two decades have been rare, and the omission of population from discussions about climate change has been particularly striking. Population growth is one of the main causes of increasing greenhouse gas (GHG) emissions around the world, and yet approaches to addressing climate change have largely ignored it (Cafaro 2012). In *This Changes Everything: Capitalism vs. The Climate*, Naomi Klein (2014), dismisses any discussion of population in a meager two sentences that is representative of this trend. After noting that the 500 million richest people on Earth are responsible for roughly half of all global GHG emissions, she makes the following remark in a footnote:

> This is why the persistent positing of population control as solution to climate change is a distraction and a moral dead end. As this research makes clear, the most significant cause of rising emissions is not the reproductive behavior of the poor but the consumer behaviors of the rich.
> (Klein 2014, p. 114 fn)

Even if the primary cause of climate change is excessive consumption by the rich, it does not follow that population reduction does not matter and cannot contribute to solving the problem. Moreover, Klein appears to assume that efforts to lower fertility rates would focus exclusively on those in developing nations. As will become apparent in the next chapter, we can (and should) pursue measures of lowering fertility rates in developed nations as well.

There is no doubt that any serious attempt to resolve our environmental problems will require radical reduction in our environmentally destructive consumption, particularly in developed nations that contribute the most to climate change and other ecological harms. As mentioned earlier, demographic momentum caused by younger populations reaching reproductive age will ensure that population growth continues for at least a generation or two further into the future, so population reduction alone cannot suffice as a solution. The pivotal question then is whether reducing consumption rates will be *enough* if population continues to rise at projected rates.

The long-term emissions reductions necessary to avoid going above a 2 °C average rise in global temperature (relative to preindustrial levels) are incredibly steep – over 5 percent per year for many nations (Raupach et al. 2014).[3] To stay below this 2 °C threshold, we must keep the concentration of GHGs in the atmosphere to 450 parts per million (ppm). If we are to stabilize our GHGs at 450 ppm, then global GHG emissions will have to decline from 2010 levels by 40–70 percent by 2050 and decline to nearly zero by 2100 (IPCC 2014, p. 20). Believing that the developed nations who must make drastic reductions can and will do so at the required pace is not only unrealistic but outright laughable. As it stands, the world is on pace for at least at least a 3 °C rise by the end of the century (Brahic 2014), and we may cross the

2 °C threshold as early as 2036 (Mann 2014). Fighting our consumption habits also requires fighting social and cultural norms. Much of the material consumption in the western world is driven by our desire for a lofty social status rather than a need for basic goods or services (Conly 2016, p. 15). Nevertheless, this status is important to many people and not something they are willing to relinquish easily.

An additional obstacle toward reducing consumption is that many nations in the world must be allowed to *increase* their rates of consumption. According to data from the World Bank (2016), 767 million people in the world were living on less than the equivalent of $1.90 per day in 2013. This level of poverty translates to very little spending and very little consumption, but those living in such circumstances struggle to survive. It would be absurd to expect or demand that these people reduce their resource consumption. Rather, they must be permitted to consume *more* so that they can escape this dehumanizing poverty. Since many of the countries with large proportions of their citizens living in extreme poverty also have high fertility rates, their increased consumption could increase environmental degradation substantially if their population growth continues.

When we take all these observations into account, we have a lot of evidence that we are reluctant to reduce our rates of consumption. Moreover, some populations must be allowed to increase their consumption in the near term. This information suggests that reducing consumption rates at the pace required to avoid severe harms from environmental degradation is extraordinarily unlikely. With respect to climate change, the prospects are absolutely dismal. At the recent Paris Climate Agreement, various goals were set to prevent a 2 °C rise in average global temperature above pre-industrial levels. To meet these goals, we must halve our carbon emissions by 2030, halve them again by 2040, and then halve them again by 2050. We must also do this while improving carbon capture technologies so much that that we are able to remove roughly five gigatons of carbon dioxide from the atmosphere each year by 2050 (Rockström et al. 2017). Making such rapid and dramatic changes to our social and economic infrastructures under the most idealized conditions would still be almost impossible; doing so is certainly impossible in the context of a rising global population.

The good news is that we do have evidence that we are willing to lower our rates of procreation. In many nations around the world, people have done it voluntarily. The fertility rate worldwide is in decline, and this trend is particularly pronounced in Western Europe, where countries like Denmark, Italy, and Germany have fertility rates far below replacement levels (CIA 2019). So while we are generally reluctant to cut back on our high-consumption behaviors, we are generally not as reluctant to reduce our rates of procreation.[4]

Our willingness to lower our fertility rates is good news because we know that procreative activities make an enormous contribution to increasing environmental degradation, especially procreation that occurs in the developed world. In one study examining the environmental consequences of having a child in the United States, the authors conclude:

> We would like all potential parents to be aware that, more than any other decision they ever make, their decision on whether or not to create a child will have the largest impact on our global environment. We conclude that the most effective way an individual can protect the global environment, and hence protect the well being of all living people, is to abstain from creating another human.
>
> (Hall et al. 1994, p. 523)

In a more recent study, Paul Murtaugh and Michael Schlax (2009) examine the carbon legacies of individuals and conclude that each new child in the United States adds about 9441 metric tons of CO_2 to an individual's carbon legacy, an amount that is roughly 5.7 times a person's lifetime emissions.[5] To offer a basis for comparison, reducing one's weekly miles driven from 231 to 155 for 80 years would only save 147 metric tons of CO_2. Thus, on their calculations, the decision to procreate will likely overshadow *all* other life choices that an American makes in an effort to reduce her individual carbon footprint. While the carbon footprint of those in other countries is not increased as much by procreating, the effect is still substantial. A new father in China has increased his carbon legacy by 4.4 times by procreating; a new mother in India has increased her carbon legacy by 2.4 times by procreating (Murtaugh and Schlax 2009, p. 18). These figures are also not static. The per capita emissions in China and India have increased significantly since this study was done and are projected to continue increasing until at least 2030 (Yeo and Evans 2015).

On a broader scale, we have compelling evidence that population growth is one of the central contributors to anthropogenic environmental stressors (Crist, Mora, and Engelman 2017; Dietz, Rosa, and York 2007; Rosa, York, and Dietz 2004). When we examine the growth of global GHG emissions, we see that they have correlated with population growth at almost a 1:1 ratio (Ryerson 2010). It does not seem possible to adequately respond to climate change without taking population seriously. The basic problem was described succinctly by Frederick Meyerson (2008) during a discussion held by the *Bulletin of Atomic Scientists*:

> Just stabilizing total emissions at current levels, while keeping pace with population growth, would require reducing global per-capita emissions by 1.2 percent each year. We haven't managed to decrease per-capita emissions by 1 percent in the last 38 years combined. The Intergovernmental Panel on Climate Change, former Vice President Al Gore, and many well-intentioned scientific, media, and activist campaigns haven't changed that fact. And because of the rapid economic growth and increased coal use in China and elsewhere, we may now be headed for higher per-capita emissions.

Historically, attempts to decrease rates of consumption have had very limited success, and future efforts may be undermined by continuing population growth. If we are serious about addressing these environmental problems, we cannot ignore population growth. The good news is that if we do decide to take population seriously, slowing the rise in population could make a substantial difference – not just in the distant future but also during this century.

Based on projections from the United Nations that estimate low, medium, and high fertility scenarios and the data we have about how population affects GHG emissions, following their projected low fertility path rather than the medium fertility path – a difference of about 0.5 births per woman – we could achieve 16–29 percent of the emissions reductions needed by 2050 to stay below the 2°C threshold (O'Neill et al. 2010). The authors of the study add, "By the end of the century, the effect of slower population growth would be even more significant, reducing total emissions from fossil fuel use by 37–41 percent across the two scenarios" (O'Neill et al. 2010, p. 17525). Even more encouragingly, some of the measures used to reduce population, such as increased funding to family planning services, are much more cost-effective in mitigating climate change than other methods (Cafaro 2012; O'Neill and Wexler 2000; Wire 2009). Moreover, for many people, increased access to family planning services may provide an easier means of decreasing their ecological footprint than reducing personal consumption. Reducing one's consumption usually requires some level of personal sacrifice, but as the data presented in earlier chapters indicates, many people desire fewer children than they ultimately have (e.g., because of unintended pregnancies). Thus, it may be in their own best interests to reduce their fertility rates.[6]

The picture painted by all these facts looks pretty clear. We cannot realistically address climate change and other environmental problems by focusing solely on reducing our rates of consumption or by focusing solely on reducing population. To be successful in responding to these problems, we need to make efforts to *both* reduce our rates of environmentally harmful consumption and reduce our population size. Returning to our equation about environmental degradation, we should focus on reducing both the factors that contribute to it.

It may feel like we have reached this conclusion too quickly, however. Before endorsing claim (9) of PRA, we need to address a significant objection. Given the seriousness of our environmental problems, many scientists and researchers are trying to develop new technologies that might provide adequate solutions to our environmental problems. Could we reasonably trust that these technologies might provide the means to reduce our consumption to the extent needed even if our population continues to grow?

The techno-optimism objection

Population growth is not a new challenge for humanity. We have dealt with rapid population growth throughout the twentieth century, and technological developments have helped us avert catastrophe. Thus, it is tempting to infer from recent history that technological advancements will once again emerge to accommodate our population growth and related environmental problems.[7] Unfortunately, as convenient as it would be if technology came to our rescue, it is unreasonable to rest our hopes entirely on technological progress.

Even if some technological optimism is justified, the rapid onset of these problems simply does not give us enough time to wait for techno-fixes to emerge. I already addressed the severity of our environmental problems in Chapter 2, but it is worth reiterating a crucial feature of them here: *they are already happening*. We are not speculating that we *might* see a rise in average global temperature in the future – the temperature rise is already happening and causing hundreds of thousands of casualties annually (DARA 2012; Global Humanitarian Forum 2009; World Health Organization 2005, 2009). We are not just viewing substantial biodiversity loss as a possibility – we are already seeing substantial biodiversity loss. It is unjustifiable to assume that technological innovation will function as a silver bullet and provide a solution to these problems in the near future.

Moreover, even if a technological fix is *possible*, it does not follow from this fact that it is likely to occur or that we should expect it to occur. It may not advance fast enough to alleviate these problems, and even if it did progress with the necessary speed, it may not actually bring us the solutions we need. Laurie Mazur and Shira Saperstein (2010) point out that the beneficial effects of technology are sometimes only realized under favorable social and economic conditions. New options for contraception and abortion can improve women's reproductive health, for example, but they "have failed to improve women's lives where underlying health, rights, and poverty issues have not also been addressed" (Mazur and Saperstein 2010, p. 12). Certainly, at this stage, we are not justified in acting as if a miracle fix is around the corner.

Furthermore, subjecting future people to such grave risk of harm is morally blameworthy even if those harms are miraculously avoided in the future. We routinely hold people accountable for engaging in actions that are unnecessarily risky even when their actions do not actually harm anyone. This is the central reason why we impose legal penalties for running red lights and driving drunk even when specific instances of those behaviors do not actually cause harm to anyone. Thus, since we have the means to minimize future people's risk of harm, it would be morally blameworthy for us not to do so.

All that said, not everyone believes that it would actually be risky to wait for a technological solution. Julian Simon (1996) claims that the discovery of a technological solution to these kinds of problems is inevitable in a free society. He argues that people – primarily because of their ability to invent and adapt – are the ultimate resource. When a resource becomes scarce, he notes that the price of this resource increases and that people gain an incentive to use this resource more effectively or develop alternatives to it. As a result, supposed shortages of resources are routinely avoided, and we should not regard natural resources as "finite in any economic sense" (Simon 1996, p. 54).

Simon's key claim is that resource scarcity plays an important role in technological advancement. He summarizes the main argument for this claim as follows:

> More people, and increased income, cause resources to become more scarce in the short run. Heightened scarcity causes prices to rise. The higher prices present opportunity and prompt inventors and entrepreneurs to search for solutions. Many fail in the search, at a cost to themselves. But in a free society, solutions are eventually found. And in the longrun *the new developments leave us better off than if the problems had not arisen.* That is, prices eventually become lower than before the increased scarcity occurred.
>
> (Simon 1996, p. 59, original emphasis)

In this manner, Simon contends that resource scarcity has a positive influence on technological progress. The rising prices caused by resource scarcity provide an economic incentive for new discoveries to be made and then put into practice. Much of Simon's *The Ultimate Resource* is an examination of how this phenomenon has occurred in the past with other resources that became scarce.

Simon also argues that a growing population increases the rate of technological progress. He first observes that improvements in productivity come from people putting their minds to use. Since these improvements originate from people, "the amount of improvement plainly depends on the number of people available to use their minds" (Simon 1996, p. 372). If other variables are held constant between two independent societies, the society with the higher population will develop more quickly because more people will be making contributions to its technological advancement.

These two argumentative threads combine to support the following conclusion: population growth drives technological advancement. Other things equal, a larger population results in the creation of a larger amount of knowledge because there are more people generating ideas and trying to put them into practice. Simultaneously, a larger population leads to faster resource depletion, resulting in an increased demand for these resources. As the prices of these resources rise, new economic opportunities emerge and provide an incentive to develop new ways of doing things (e.g., using the resource more efficiently, finding new sources of the resource, developing alternatives to the resource). In a free society, solutions are eventually found, and in the end, the prices of the resources end up being lower than they would have been if the original scarcity had never arisen. In this manner, we ultimately end up being better off for having endured this (temporary) resource scarcity. If one is persuaded by this line of reasoning, then significant restraints on population growth may seem not just unnecessary but *detrimental*, since reduced population growth will hinder our technological advancement.

Simon is right to point out that this trend has happened many times in human history, but this observation is not enough to sustain his argument. It simply has too many shortcomings. First, his understanding of "better off" is purely economic: being better off simply means that we are in an economically superior position. But that does not always mean that we are better off overall. If resource scarcity (even if temporary) is so pronounced and devastating that it results in the deaths of millions of people and the severe suffering of many million more, in what sense are we "better off" after this scarcity concludes? Even if we are better off in some narrowly economic sense, we may well be worse off in terms of aggregate human welfare, a measurement that seems to have much greater moral significance. This point is perhaps most poignant with respect to climate change.

Climate change can be interpreted as the scarcity of a resource – namely, the available carbon sinks on Earth. We have too few carbon sinks to accommodate our GHG production. This scarcity has created plenty of incentive to develop means of increasing our available carbon sinks. Some have investigated the possibility of geoengineering the atmosphere to aid climate change mitigation, but all of them carry significant risks and uncertainties. Some also do not appear economically viable.[8] Herein lies a second problem with Simon's reasoning: he assumes that technological solutions to problems of scarcity will always be found within a viable timeframe. That has been the case many times in the past, but what justifies assuming that it will always happen regardless of the circumstances? When the moral stakes are this high, it is both irrational and morally unjustifiable to exacerbate a problem in the *hope* that it will motivate people to develop a solution.

Climate change in particular is an environmental problem that Simon does not properly address. In fact, some of his remarks about it are outright dismissive:

> Given the history of such environmental scares – over all of human history – my guess is that global warming is likely to be simply another transient concern, barely worthy of consideration ten years from now should I then be writing again of these issues.
>
> (Simon 1996, p. 266)

This particular remark did not age well and reveals a rather large blind spot in his argument. Climate change is still a pressing concern, and a technological solution to climate change is nowhere to be found. This seems to be a straightforward counterexample to the claim that all problems of resource scarcity will be resolved through technological advancement.

Simon was similarly mistaken about biodiversity loss. He claims that he and Aaron Wildavsky "documented the complete absence of evidence for the claim that species extinction is going up rapidly, or even going up at all" in the mid-1980s and that no one disputed their documents or "adduced any new evidence since then that would demonstrate rapid species extinction" (Simon 1996, p. 450). Whatever the state of conservation biology 30 years ago, the studies I cited in Chapter 2 provide ample evidence that we are experiencing rapid species extinctions. The consensus among conservation biologists on this point is overwhelming. Much like climate change, the problem is on our doorstep, and technological progress has not been able to solve it.[9] Technological advances can certainly play a role in our efforts curb our environmental impact, but we cannot continue with business as usual under the expectation that technology will suddenly solve all our problems. Such a path would be both foolish and morally reprehensible.

Concluding the population reduction argument

We are now finally in position to reach the end of PRA. The preceding section provides firm justification for endorsing the argument's final premise:

9   There is no feasible way to reduce the rate of environmental degradation per person enough to adequately respond to our environmental problems if our population size remains at its current size or continues to grow.

With this premise secured, we can connect claims (7), (8), and (9) to reach the following conclusion:

10   We morally ought to both reduce our rates of environmental degradation per person and reduce our current population size.

The only reasonable means of responding to the impacts of environmental degradation is to address *both* of its contributing factors. We must pursue measures to reduce the ecologically destructive behaviors that give lead to high rates of environmental degradation per person, and we must also do what we can to limit our population size. Since we should do both these things, we can logically infer the final claim of PRA:

11   We morally ought to reduce our current population size.

Of course, the duty to reduce our current population size must be undertaken over the long term because reducing our population below its current level will not be achievable within the next two generations. Before reduction can occur, we must stabilize global population and then begin the descent from that peak – whatever that number happens to be.

The quest for population stabilization brings us to the next stage of our inquiry. How could we ensure that human population peaks at a lower number than 10.9 billion people? In the next chapter, I will assess the policy measures that could be permissibly used to achieve this goal.[10]

Notes

1 For an overview of how the IPAT equation has been employed in the past, see Chertow (2000).
2 This surprising phenomenon, known as the Jevons paradox, is one of the most widely known paradoxes in environmental economics (York 2006).
3 A temperature rise of 2 °C is usually presented as the policy goal, but a recent report by the IPCC (2018) suggests that the impacts at 1.5 °C would still be quite significant, though obviously not as bad as a temperature rise of 2 °C or higher.
4 Conly (2016) makes the same observation (pp. 17–18).
5 They calculate carbon legacy on the assumption that "a person is responsible for the emissions of his descendants, weighted by their relatedness to him. For a descendant that is n generations removed from the focal individual, the weight is $(1/2)n$" (Murtaugh and Schlax 2009, p. 14). So a person is responsible for one-half the emissions of her children, one-fourth the emissions of her grandchildren, one-eighth the emissions of her great-grandchildren, and so on.
6 Hickey, Rieder, and Earl (2016) make this same point and also add that preference-adjusting interventions (which I will discuss in the next chapter) could make people want fewer children (p. 870).
7 For a recent discussion and endorsement of this claim, see Pearce (2010, pp. 204–208).
8 See Boyd (2008) for a brief appraisal of different geoengineering schemes.
9 In theory, efforts to create synthetic organisms could enable us to create new organisms that fill the same ecological role as species that have gone extinct or even to genuinely resurrect extinct species. But these efforts are nowhere near coming to fruition to the extent that would be required to genuinely avert biodiversity loss or recover from it. It was only quite recently that we even managed to create the first synthetic cell (Gibson et al. 2010).
10 Significant portions of this chapter are derived from chapter 5 of my doctoral dissertation. See Hedberg (2017).

References

Boyd, Philip W. 2008. "Ranking Geo-Engineering Schemes." *Nature Geoscience* 1: 722–724.
Brahic, Catherine. 2014. "World on Track for Worst-Case Warming Scenario." *New Scientist*. www.newscientist.com/article/dn26243-world-on-track-for-worst-case-warming-scenario/. Accessed December 7, 2019.
Cafaro, Phil. 2012. "Climate Ethics and Population Policy." *Wiley Interdisciplinary Reviews: Climate Change* 3, no. 1: 65–81.
Chertow, Marian. 2000. "The IPAT Equation and Its Variants." *Journal of Industrial Ecology* 4, no. 4: 13–29.
CIA (Central Intelligence Agency). 2019. "Country Comparison: Total Fertility Rate." *The World Factbook*. www.cia.gov/library/publications/the-world-factbook/rankorder/2127rank.html. Accessed December 7, 2019.
Conly, Sarah. 2016. *One Child: Do We Have a Right to Have More?* Oxford: Oxford University Press.
Crist, Eileen, Camilo Mora, and Robert Engelman. 2017. "The Interaction of Human Population, Food Production, and Biodiversity Protection." *Science* 356, no. 6335: 260–264.

The moral duty to halt population growth

DARA (Development Assistance Research Associates). 2012. *Climate Vulnerability Monitor 2nd Edition: A Guide to the Cold Calculus of a Hot Planet*. Madrid: DARA and the Climate Vulnerable Forum. http://daraint.org/climate-vulnerability-monitor/climate-vulnerability-monitor-2012/report/. Accessed December 6, 2019.

Dietz, Thomas, Eugene A. Rosa, Richard York. 2007. "Driving the Human Ecological Footprint." *Frontiers in Ecology and the Environment* 5, no. 1: 13–18.

Gibson, Daniel, John Glass, Carole Lartigue, Vladimir Noskov, Ray-Yuan Chuang, Mikkel Algire, Gwynedd Benders, Michael Montague, Li Ma, Monzia Moodie, Chuck Merryman, Sanjay Vashee, Radha Krishnakumar, Nacyra Assad-Garcia, Cynthia Andrews-Pfannkoch, Evgeniya Denisova1, Lei Young, Zhi-Qing Qi, Thomas Segall-Shapiro, Christopher Calvey, Prashanth Parmar, Clyde Hutchison III, Hamilton Smith, J. Craig Venter. 2010. "Creation of a Bacterial Cell Controlled by a Chemically Synthesized Genome." *Science* 329, no. 5987: 52–56.

Global Humanitarian Forum. 2009. *Climate Change: The Anatomy of a Silent Crisis*. www.ghf-ge.org/human-impact-report.pdf. Accessed December 6, 2019.

Hall, Charles, R. Gil Pontius, Jr., Lisa Coleman, and Jae-Young Ko. 1994. "The Environmental Consequences of Having a Baby in the United States." *Population and Environment: A Journal of Interdisciplinary Studies* 15, no. 6: 505–524.

Hedberg, Trevor. 2017. "Population, Consumption, and Procreation: Ethical Implications for Humanity's Future." Ph.D. dissertation, Department of Philosophy, University of Tennessee.

Hickey, Colin, Travis Rieder, and Jake Earl. 2016. "Population Engineering and the Fight against Climate Change." *Social Theory and Practice* 42, no. 4: 845–870.

IPCC (Intergovernmental Panel on Climate Change). 2014. "Summary for Policy Makers." *Climate Change 2014 Synthesis Report*. www.ipcc.ch/report/ar5/syr/. Accessed December 7, 2019.

IPCC (Intergovernmental Panel on Climate Change). 2018. *Global Warming of 1.5 °C: Summary for Policy Makers*, edited by Valérie Masson-Delmotte, et al. Switzerland: IPCC. https://report.ipcc.ch/sr15/pdf/sr15_spm_final.pdf. Accessed December 7, 2019.

Klein, Naomi. 2014. *This Changes Everything: Capitalism vs. The Climate*. New York: Simon & Schuster.

Mann, Michael. 2014. "Earth Will Cross the Climate Danger Threshold by 2036." *Scientific American*. www.scientificamerican.com/article/earth-will-cross-the-climate-danger-threshold-by-2036/. Accessed December 7, 2019.

Mazur, Laurie, and Shira Saperstein. 2010. "Beware the Techno-Fix." In *A Pivotal Moment: Population, Justice and the Environmental Challenge*, 2nd ed., edited by Laurie Mazur, 12–13. Washington, D.C.: Island Press.

Meyerson, Frederick. 2008. "Population Growth Is Easier to Manage than Per Capita Emissions." *Bulletin of the Atomic Scientists*. http://thebulletin.org/population-and-climate-change/population-growth-easier-manage-capita-emissions. Accessed December 7, 2019.

Murtaugh, Paul, and Michael Schlax. 2009. "Reproduction and the Carbon Legacies of Individuals." *Global Environmental Change* 19, no. 1: 14–20.

O'Neill, Brian, Michael Dalton, Regina Fuchs, Leiwen Jiang, Shonali Pachuri, and Katarina Zigova. 2010. "Global Demographic Trends and Future Carbon Emissions." *Proceedings of the National Academy of Sciences* 107, no. 41: 17521–17526.

O'Neill, Brian, and Lee Wexler. 2000. "The Greenhouse Externality to Childbearing: A Sensitivity Analysis." *Climatic Change* 47, no. 3: 283–324.

Pearce, Fred. 2010. *The Coming Population Crash and Our Planet's Surprising Future*. Boston: Beacon Press.

Raupach, Michael, Steven Davis, Glen Peters, Robbie Andrew, Josep Canadell, Philippe Ciais, Pierre Friedlingstein, Frank Jotzo, Detlef van Vuuren, and Corinne Le Quéré. 2014. "Sharing a Quota on Cumulative Carbon Emissions." *Nature Climate Change* 4, no. 10:873–879.

Rockström, Johan, Owen Gaffney, Joeri Rogelj, Malte Meinshausen, Nebojsa Nakicenovic, Hans Joachim Schellnhuber. 2017. "A Roadmap for Rapid Decarbonization." *Science* 355, no. 6331: 1269–1271.

Rosa, Eugene A., Richard York, and Thomas Dietz. 2004. "Tracking Anthropogenic Drivers of Ecological Impacts." *AMBIO: A Journal of the Human Environment* 33, no. 8: 509–512.

Ryerson, William. 2010. "Population: The Multiplier of Everything Else." In *The Post Carbon Reader: Managing the 21st Century's Sustainability Crises*, eds. Richard Heinberg and Daniel Lerch, 153–174. Healdsburg, CA: Watershed Media.

Simon, Julian L. 1996. *The Ultimate Resource 2*. Princeton: Princeton University Press.

UN Department of Economic and Social Affairs, Population Division. 2019. *World Population Prospects 2019: Volume I: Comprehensive Tables*. https://population.un.org/wpp/Publications/Files/WPP2019_Volume-I_Comprehensive-Tables.pdf. Accessed December 5, 2019.

Wire, Thomas. 2009. *Fewer Emitters, Lower Emissions, Less Cost: Reducing Future Carbon Emissions by Investing in Family Planning: A Cost/Benefit Analysis*. London: London School of Economics.

World Bank. 2016. *Poverty and Shared Prosperity 2016: Taking on Inequality*. Washington, D.C.: International Bank for Reconstruction and Development/The World Bank. https://openknowledge.worldbank.org/bitstream/handle/10986/25078/9781464809583.pdf. Accessed December 7, 2019.

World Health Organization. 2005. *Climate and health: Fact sheet, July 2005*. www.who.int/globalchange/news/fsclimandhealth/en/index.html. Accessed December 6, 2019.

World Health Organization. 2009. *Global Health Risks: Mortality and burden of disease attributable to selected major risks*. Geneva: WHO Press. www.who.int/healthinfo/global_burden_disease/GlobalHealthRisks_report_full.pdf. Accessed December 6, 2019.

Yeo, Sophie, and Simon Evans. 2015. "Analysis: India's Climate Pledge Suggests Significant Emissions Growth up to 2030." *Paris Summit 2015*. www.carbonbrief.org/indias-indc. Accessed December 7, 2019.

York, Richard. 2006. "Ecological Paradoxes: William Stanley Jevons and the Paperless Office." *Human Ecology Review* 13, no. 2: 143–147.

5 Policies that promote smaller families

If we accept the conclusion of the Population Reduction Argument, then we have a moral duty to pursue actions that reduce global population. In the short term, that means trying to hasten the deceleration of population growth and ensure that the population peaks sooner and at a lower number than what is currently projected. In Chapter 2, I mentioned that the United Nations Department of Economic and Social Affairs (2019) estimates that the global population will be about 10.9 billion people in 2100. But there is some good news about this estimate. This figure originates from the medium variant of their projections, and there is a much wider range of possible outcomes. Their upper 95 percent prediction interval points to a global population of almost 12.7 billion people in 2100, and their lower 95 percent prediction interval estimates the global population at just above 9.4 billion in 2100. That means that the medium-variant estimate of 10.9 billion is not fixed: decisions that we make during the next 80 years could alter the size of the global population in 2100 by several billion people. Now we can consider what should be done to reduce the rate of population growth over the next few generations and provide the groundwork for population reduction in the generations that follow.

Part of population growth has been caused by an increase in the life expectancy of people around the world, particularly in Africa (Johnson 2016; Kweifio-Okai and Holder 2016). While reproductive rates have declined overall, decreases in the rate of population growth have been muted because of the decrease in death rates. However, since the increases in life expectancy are a result of better medical care and a significant reduction in human misery, we should not deliberately aim to lower life expectancy.[1] A far better way to reduce the population is to bring fewer people into existence.

The concern about trying to lower fertility is that doing so will involve morally problematic coercion. These worries are not unfounded: the implementation of policies in China, India, and Peru aimed at reducing fertility rates resulted in forced abortions and sterilizations (Alvarado and Echegaray 2010; Mosher 2008, chs. 3 and 5). These practices are widely regarded as human rights violations and thought morally indefensible. Whatever we do in response to the population problem, we must avoid a repeat of these

64 *Ethics, policy, and obligations*

inhuman practices. We need to slow population growth as swiftly as we can while also respecting people's personal freedoms. Thus, we should attempt to implement the least coercive set of population policies possible that will still address the problem effectively.

In the remainder of the chapter, I will examine various policy measures to see which ones could help us make progress in halting population growth without being objectionably coercive. I divide these into three broad categories: autonomy-enhancing measures, semi-coercive measures, and severely coercive measures. Overall, I will argue that autonomy-enhancing measures and semi-coercive measures are worth pursuing and that severely coercive measures should be avoided.

Autonomy-enhancing measures

In this section, I will focus on the means of lowering fertility rates that involve *increasing* people's autonomy. The first of these is increasing access to contraception and family planning services. Doing so increases reproductive autonomy by giving prospective parents greater control over their reproductive choices. Because increasing the availability of these services is both effective and non-coercive, this strategy for lowering fertility enjoys near universal support among those who have addressed our rising population size (e.g., Cafaro 2012; Hickey, Rieder, and Earl 2016; Kukla 2016; Mazur 2010; Ryerson 2010). Since improving access to contraception and family planning is also rather cheap, this means of lowering fertility generates a significant environmental benefit while enhancing procreative freedom and remaining cost-effective (Bongaarts and Sinding 2011).

Much progress could be made in lowering fertility rates if we were to provide contraception to all who have an unmet need for it. Worldwide, only 56 percent of married women between the ages of 15 and 49 use modern methods of contraception, and in Africa, this figure dips to 30 percent (Population Reference Bureau 2016). About 12 percent of the women in the world want to delay or prevent childrearing but are not using any methods of contraception; in the developing world, this figure rises to 22 percent (UN Department of Economic and Social Affairs 2015). These figures highlight how increased funding for family planning programs could make a significant difference in slowing population growth. Just meeting the contraceptive needs of Africa could decrease the global population in 2030 by as much as one billion (Ford 2016).

Of course, we also have evidence that increased access to family planning services is not enough. Globally, 40 percent of pregnancies are unplanned (Sedgh, Singh, and Hussain 2014), and a significant portion of pregnancies remain unplanned even in parts of the world where contraception is readily available. Thus, measures must be taken to improve people's awareness of how to use contraceptives effectively and the risks associated with not using them. The most straightforward way to accomplish this feat is to improve the

Policies that promote smaller families 65

availability and quality of sex education. What this entails may vary from nation to nation according to their educational system, but whatever education is provided should include information on how to use contraception effectively. Abstinence-only programs, which promote abstinence until marriage and do not cover contraceptive use, have been in place in certain regions in the United States for decades. These programs have consistently received federal funding during the last 20 years, but they have proven utterly ineffective in reducing rates of unintended pregnancies and sexually transmitted infections compared to comprehensive sex education (Advocates for Youth 2007; Breuner and Mattson 2016; Stranger-Hall and Hall 2011). Of course, the United States is no model for how to educate the youth about sex: only 29 states mandate sex education of any kind, and only 20 states require sex education that includes content related to the use of contraception (Guttmacher Institute 2019). Given these facts, we should not be surprised that the United States has the highest rates of unintended pregnancy in the developed world.

Comprehensive sex education improves people's autonomy by making them more aware of both the choices available to them and the consequences of those choices and also results in lower fertility through fewer unintended pregnancies. Pursuing gender justice in a broader sense yields similar results. Fertility rates drop significantly when women are not blocked by various social and cultural factors from exercising control over their reproductive decisions (Crist 2019, ch. 8; Roudi, Fahimi, and Kent 2007). Countering patriarchal norms and other influences that strip women of their procreative autonomy can play a significant role in reversing population growth.

The case for improving access to contraception, improving sex education, and pursuing gender equity is compelling. Doing so will enhance people's freedom (especially the reproductive freedom of women in the developing world) and improve their quality of life (since they will have fewer unwanted children) while also lowering fertility rates. These policies have, as Rebecca Kukla (2016) puts it, "no significant moral downside" (p. 845). The real question is whether or not these measures would be enough to effectively respond to population growth. Suppose we give everyone in the world ready access to contraception, improve sex education significantly, and make substantial gains in gender equity across the world. Under such circumstances, would the population problem be solved?

Unfortunately, it is difficult to know what the precise effects of meeting these conditions would be. It would certainly be convenient if these changes alone solved the problem: then we would not need to worry about answering the more difficult ethical questions about coercive policies. Some do genuinely believe that improved access to contraception and increased awareness of how to use it effectively will solve the problem. In the introduction to her edited volume on the population problem, Laurie Mazur (2010) states, "It is not necessary to control anyone to slow population growth: Birthrates come

66 *Ethics, policy, and obligations*

down where individuals have the means and power to make their own reproductive choices" (p. 16).

Despite Mazur's optimism, I think it is naïve to believe that improvements in access to family planning, sex education, and gender justice would be sufficient to solve the problem for three reasons. First, the data on population suggests that unmet contraceptive needs are not the only major contributor to population size. Consider a few examples based on recent population data (Population Reference Bureau 2016). In Morocco, 57 percent of the married women aged 15–49 use modern contraceptive methods, and the fertility rate is 2.4; in Malawi, 58 percent of married women in this age range use modern contraceptive methods, and the fertility rate is 4.4. In Libya, only 20 percent of these women use modern contraceptive methods, but the fertility rate is only 2.4. Women of the same demographic in Senegal use modern contraceptives at almost the same rate as those in Libya (21 percent), and yet the fertility rate in Senegal is 5.0. Something other than contraceptive access must be playing a large role in fertility rates, and a plausible culprit is the family size desired by the country's citizens (Ryerson 2012, pp. 241–243). Economist Lant Pritchett (1994) went so far as to claim that the desire for children was the primary determinant of fertility rates and that "contraceptive access (or cost) or family planning effort more generally is not a dominant, or typically even a major, factor in determining fertility differences" (p. 39).

Additionally, even under a best-case scenario where we implement these measures to increase access to contraception, improve sex education, and pursue gender justice, we will not reduce fertility rates quickly enough to deal adequately with the environmental problems we now face. According to recent demographic models, the human population in this scenario would still closely approximate the nearly 11 billion that we will otherwise have on Earth in 2100; substantial reductions in the population are unlikely to occur until the following century (Bradshaw and Brook 2014, pp. 16611–16612). We must take significant action this century to avert the most severe climate change and biodiversity loss, so these actions, though important, will not be enough by themselves.

Finally, because developed nations have the largest per capita ecological footprints, they are the places in the world where population reduction would be most beneficial. While these countries often already have fertility rates lower than replacement levels, decreasing the fertility rate a bit more in developed nations could make a much larger difference to our overall environmental impact than greater reductions in developing nations. Hickey, Rieder, and Earl (2016) offer a succinct encapsulation of this reasoning:

> While reducing fertility in developing nations is important, since their per capita GHG emissions are projected to increase significantly (and should be allowed to do so) over the next several decades, it is not nearly as critical as near-term reductions in the numbers of the world's wealthy. Although it would be difficult to lower the fertility rate in the United

States from 1.9 to, say, 1.4, such a reduction would have a massive impact on both near-term and long-term global GHG emissions – much more even than proportionally larger fertility reductions in sub-Saharan Africa.

(pp. 855–856)

Given the gravity of the problem and the need to act quickly, we must consider the ways in which we can lower fertility rates in the developed world even more, particularly in countries like the United States and the United Kingdom where the per capita ecological footprint is high and fertility rates are still close to two children per woman (CIA 2019). In most cases, citizens in these nations already have access to family planning services and do not confront the same issues with gender justice that exist elsewhere.[2] Thus, we need to consider some other measures to aid fertility reduction in these nations. However, before discussing some alternative strategies for lowering fertility rates, I must address an important question about what increased access to family planning services would entail.

Does the population problem justify a liberal abortion policy?

Readers may suspect that increasing access to family planning services entails that women should have the right to abort unwanted pregnancies and that this service should be readily available to them. There is no question that increasing access to abortion services increases women's reproductive autonomy, so it would be an autonomy-enhancing measure. But abortion is a more controversial procedure than the use of contraception. I do believe that women should have the ability to obtain abortions during the first trimester of pregnancy (and in some circumstances later in the pregnancy), but my reasons for holding this view are not tied to the problem of population growth.

To reiterate an earlier point, we should not strive to lower population by causing existing people to die. One important implication of that principle is that it would be wrong to abort a fetus to reduce population *if the fetus is a person*. In this context, a person is an entity with a moral status equivalent to that of an adult human being. If a fetus is a person from the moment of conception, then abortion will be morally equivalent to murder, and so it will not be a permissible means lowering fertility rates. Thus, whether we should increase access to abortion as part of increasing access to family planning services hinges significantly on whether the fetus is a person.

The morality of abortion and the issues concerning fetal personhood are too complex to discuss at length here, but I will make a few general remarks to clarify my position. As an important initial observation, even assuming that the fetus is a person from the moment of conception, there are compelling arguments that abortion remains permissible in certain circumstances (Thomson 1971). One such circumstance is when pregnancy occurs as a result of rape. When a women is impregnated against her will, it is unfair to demand that she endure the burdens of pregnancy, and while it is unfortunate that the fetus will die as a result, we do not typically require people to endure substantial burdens to save the lives of others when they are not responsible for the other person being in life-threatening circumstances. A woman completing a pregnancy that results from rape goes well beyond what morality requires. The other commonly recognized exception is when the continuation of pregnancy endangers the mother's life. In this case, the mother's right to self-defense justifies her ending the fetus's life to preserve her own.

As a second general point, it is implausible to regard a fetus as being a person from the moment of conception. As Mary Anne Warren (1973) argues, an early term fetus does not have *any* of the qualities that we typically associate with personhood. She identifies the following features as being typical components of personhood: consciousness and the capacity to feel pain, the ability to reason, engagement in self-motivated activity, the ability to communicate, and the presence of self-concepts and self-awareness (Warren 1973, p. 55). An early term fetus does not have *any* of these features. An entity probably does not need all of them to be a person, but it surely needs *at least one* of them. An early term fetus does not possess any level of conscious awareness, which seems like a prerequisite for possessing the other features of personhood. Thus, at least early in the pregnancy, its moral status should be similar to that of other living things that lack the capacity for consciousness (e.g., plants).

The moral picture gets more complicated as pregnancy progresses, however. Sometime during the pregnancy, the fetus becomes sentient, which means that it acquires the capacity to feel pleasure and pain. The precise time at which the fetus becomes sentient is a subject of controversy. Some have placed the threshold for sentience about 30 weeks after conception (Lee et al. 2005; Tawia 1992) while others contend that the fetus can feel pain closer to 20 weeks after conception (Grossu 2017). Regardless, once the fetus becomes sentient, it acquires an interest in avoiding pain, and this new capacity results in an elevation in its moral status. The fetus now has one feature of personhood and appears to be in the same moral category as a wide assortment of nonhuman animals. Past this point, the justification for an abortion must be significantly stronger than the justification offered for aborting a non-sentient fetus. In practice, recognizing the significance of sentience might result in a policy of permitting the abortion of fetuses during the first trimester (when the fetus is clearly not sentient) and only allowing abortions in exceptional circumstances after the first trimester (Sumner 1981, ch. 4). Such circumstances could include, for instance, a threat to the mother's health or the discovery of significant genetic defects in the fetus.[3]

Admittedly, there is one significant objection to this approach to the morality of abortion. In one of the most widely anthologized papers on abortion, Don Marquis (1989) argues that abortion is wrong because it deprives the fetus of future experiences. Marquis believes this is the same reason killing an adult human being is wrong, so he views abortion as being just as wrong as murdering an adult human being. At first glance, this argument appears to provide a reason to oppose abortion that does not rely on the claim that the fetus is a person. However, Marquis's argument only establishes that the fetus has a valuable good that it can lose – namely, its future. This fact alone does not establish that the fetus is the kind of entity that has a right to its future or that there is anything morally wrong with depriving it of this good (Sinnott-Armstrong 1999). For it to have this kind of moral status, it would have to be a person (or something similar). So Marquis does not actually succeed in bypassing the issue of fetal personhood.[4]

I believe that increasing access to family planning services should also entail giving women greater access to abortion services during (at least) the first trimester, but this position does not result from thinking that the imperative to reduce population automatically warrants allowing more women to receive abortions. Rather, it follows from my views about the moral status of the fetus. Even if someone held all my views regarding the need to reduce population, they could reach a different conclusion about whether we should make it easier for women to obtain abortions if they held a different view about the moral status of fetuses.

Semi-coercive measures

Autonomy-enhancing measures are the least controversial way to lower fertility because they give people *greater* reproductive freedom. I now want to consider a range of measures that are neither autonomy-enhancing nor severely coercive. I start with two strategies that are widely used in other contexts and generally regarded as permissible: preference adjustment and incentivization. In this context, preference adjustment involves trying to lower fertility rates by changing cultural norms or individual desires, and incentivization involves providing incentives for people to have fewer children. Incentives can be either positive or negative. Positive incentives are those that provide benefits to those who have few children, and negative incentives are those that impose penalties on people who have too many children. Both these strategies are often regarded as permissible in other contexts and not thought to constitute rights violations. We use them to protect people from harm and advance public interests. Some examples include influencing people's dietary habits and sexual behavior to lower public health costs, creating incentives to make certain careers more attractive, and encouraging certain behaviors that generally make people safer (e.g., wearing seatbelts). These practices are widely accepted but not thought to violate anyone's rights or prevent them from living autonomously (Hickey, Rieder, and Earl 2016, p. 857). Given the general acceptance of these other practices, we ought to consider the effectiveness of them in the realm of procreation.

The primary means of adjusting people's preferences would be through the use of mass media – radio, television, poster campaigns, billboards, advertising on popular online video media (e.g., YouTube, Twitch, Hulu), and so on.

Sometimes, preference adjustment takes the form of rational persuasion, which involves objective presentation of factual information. Other times, the persuasion is more subtle and involves trying to change behavior through tactics like appeals to emotion, celebrity endorsements, or presentation through a narrative. Although some might worry that these latter strategies constitute undesirable manipulation, this objection is weak. These strategies are already widely employed in a variety of these contexts without causing controversy, and they do not need to present false information or to be undertaken covertly. Moreover, some cultures are dominated by pronatalist values. In these cultures, it is normal and expected that people will have children. Preference-adjusting campaigns could serve to counter this pronatalism and make it more socially acceptable for people to remain childless. In doing so, they would enhance individuals' autonomy by alleviating the social and cultural pressure to have children (Hickey, Rieder, and Earl 2016, p. 860).

Preference-adjusting interventions have been implemented before, and they have proven effective. Television shows that promoted family planning and small family size aired in Mexico during the 1970s and 1980s, and similar programs were later launched in India. Kenya and Tanzania promoted the same values through radio programs. In all these cases, the launch of these media programs was followed by a decline in fertility rates and an increase in contraceptive use (Ryerson 2012, pp. 244–248). These programs often shifted their audience's beliefs about the acceptability of family planning and their perceptions of family size. As a result, viewers became more likely to use contraception, delay childbearing, and have fewer children (Rogers et al. 1999; Singhal and Rogers 1989). William Ryerson (2012) estimates that expenses of $35 million per year would be sufficient to fund similar programs in all the world's major developing countries (p. 448). That financial estimate might be too optimistic, but it is clear that media-driven preference adjustment could be an effective policy tool with respect to reducing family size.

The use of incentives would be trickier because some incentives creep uncomfortably close to the threshold of unacceptable coercion. Negative incentives, such as severe fines or increased hospital delivery fees, may be indistinguishable from outright coercion when they are imposed on people who are in financially precarious circumstances. Moreover, some negative incentives in the past have been imposed in ways that are clearly objectionable. China's incentive-oriented policies often pressured mothers to abortion and infanticide (Hesketh and Xing 1997; Thomas 1995, p. 10), and India's incentives – clothing, electronics, and monetary payments designed to encourage sterilization or delayed childbearing – exploited the low literacy rate among the poor to sterilize thousands without their informed consent (Repetto 1968). Given their morally repugnant nature, these incentivizing strategies must be avoided. At the same time, a blanket dismissal of incentivization would be too hasty. Incentives can be effective in lowering fertility rates despite differences in cultural norms and resource availability (Heil, Gaalema, and Herrmann 2012), so they could be worth using if we could minimize the extent to which they would lead to injustice.[5]

Fortunately, some measures can be taken to reduce the risk that incentives will be exploitative or objectionably coercive (Hickey, Rieder, and Earl 2016). First, we can be transparent about the political goals behind the incentives, the methods that are used, and the actual outcomes that result from them. Second, we can restrict payment for incentives to the actual would-be procreators. In China, local and regional officials were offered incentives to reduce the fertility rates of their constituents (Hesketh and Xing 1997; Thomas 1995, p. 7), and in India, incentives were offered to various intermediaries to encourage other people to be sterilized (Repetto 1968, p. 13). These practices increase the risk that would-be procreators will be pressured by others into altering their reproductive behavior rather than it resulting from their own voluntary decisions. Third, we can take precautions to try to reduce the impact of incentive-based interventions on vulnerable groups. One means of doing this would be to direct positive incentives toward these vulnerable groups and reserve negative incentives for other, less vulnerable groups. For example, we could offer cash payments and tax breaks to the poor and levy fines against the wealthy (Hickey, Rieder, and Earl 2016, p. 868). On such a scheme, the poor would not be made worse off by a decision to have a large family; they would simply have to forego benefits that they would otherwise be able to obtain.

Directing positive incentives toward vulnerable groups has the added advantage of avoiding scenarios where children are heavily disadvantaged by the actions of their parents, a worry raised by Cripps (2016, p. 382). If the poor were subjected to fines, then there might be circumstances where a child's welfare is threatened because the parents are heavily fined for giving birth to the child. Such scenarios seem deeply unjust because the child who is born and that child's siblings will be the ones most adversely affected by the fines, and these children have no control over the circumstances of their birth.

Another incentivization strategy worth considering is the use of procreation entitlements (Bognar 2019; de la Croix and Gosseries 2009). Suppose that we want to lower fertility rates in the United States to about 1.5 births per woman. We might grant everyone in the United States a sellable entitlement of 0.75 children. Now imagine that this couple has a child, so each of their individual entitlements drops to 0.25 children. These new parents would now have a choice. If they do not want to have any more children, then they could put their entitlements on an open market and sell them. If they wanted to have more biological children, then they could purchase additional entitlements on this marketplace. In effect, this would create an additional economic incentive for people to have fewer children, since they would have to both forego the income they could get from selling their entitlements and purchase additional entitlements if they want a large family. This entitlement scheme could potentially be used in conjunction with other economic incentives. Perhaps a penalty for having more biological children than a person's entitlements allowed would be to forego certain tax exemptions that a person normally qualifies for by having dependents.

Undoubtedly, a procreative entitlement scheme would confront some significant logistical challenges. One question would be how we determine precisely what number of entitlements a person has by default. Perhaps the entitlements would need to be context sensitive depending on the ecological footprint of the country's citizens. An average person living in Niger, for example, has an ecological footprint that is less than one-fifth of the ecological footprint of an average person living in the United States (Global Footprint Network 2019). It appears unreasonable for those living in Niger to have the same entitlement scheme as people in the United States. Such a policy would disproportionately restrict the freedom of people who are contributing relatively little to the environmental problems that motivate the policy.[6] Perhaps those in the United States should only have an entitlement of 0.5 children per person whereas those in Niger should have an entitlement of one child per person. The specific numbers are debatable, of course, but the point is that some variability in the limits on reproduction is appropriate given the radical difference in ecological impact that the citizens in these countries have. In any case, the bigger challenge would probably be one of political feasibility. It is hard to envision this particular proposal gaining traction in the immediate future, no matter how much we fine-tune its specifics. That may change in the future, but in the short term, this measure looks unlikely to make our list of viable policy responses.

One final semi-coercive measure to consider is the implementation of mandatory long-term contraception (Bognar 2019, pp. 320–324). In the status quo, contraception is something that must be purchased and intentionally used in order to prevent pregnancy. Procreation is the default result of sexual activity. Levonorgestrel and etonogestrel contraceptive implants already exist, though they can only be used by women and last for only a few years. It is not too farfetched to imagine longer-lasting versions of these contraceptives and versions that are useable by men. If these contraceptives were effective and lasted for a long enough period of time, they could virtually eliminate unintentional pregnancies. Pregnancy would, in Greg Bognar's words, "entirely be a matter of choice, rather than chance" (p. 322). In this respect, the use of long-term contraception appears to enhance procreative liberty rather than restricting it. Nonetheless, I consider this measure semi-coercive because it would require an initial infringement on autonomy to achieve this long-term benefit to procreative liberty. That infringement may be acceptable, however. As Bognar (2019) mentions, we do not typically regard certain public health initiatives, such as mandatory immunizations, to be objectionable even though they are coercive to some degree.

The moral case in favor of mandatory long-term contraception is fairly strong, but since the contraceptive implants available are currently limited in their longevity and cannot be used by men, implementing this strategy for responding to population growth is not yet viable. Even so, should it become a viable option in the future, it may be worth considering – especially if other attempts to reduce population growth are insufficiently successful.

We can now draw a few conclusions from the discussion so far. First, autonomy-enhancing measures could make a significant impact on reducing population growth while also increasing people's reproductive autonomy. These measures include making contraception and family planning services more widely available, improving sex education, and promoting gender equity. All of these strategies should be pursued to the fullest extent possible. Second, preference adjusting interventions – mainly done through use of mass media – should play a role in countering pronatalist values and encouraging people think more critically about their procreative choices. So long as they are not done in deceptive ways, these strategies will be morally analogous to a variety of other preference-adjusting interventions that we routinely permit. Third, the implementation of incentive-based schemes to lower fertility rates would be more morally treacherous than most other options and require significant efforts to guard against injustice. For these reasons, the use of these schemes should only be considered after other options have been exhausted.

We also need to recognize that the best strategies for responding to population growth will vary depending on the context. Autonomy-enhancing measures and preference-adjustment interventions should be the main strategies for reducing fertility in the developing world. If incentivization schemes are implemented, they should be restricted to the developed world for the moment. Incentivizing measures are the most coercive of those under consideration, and it is morally appropriate to exert more pressure on wealthier individuals to lower fertility rates than on others (Hickey, Rieder, and Earl 2016, p. 868). Moreover, those who are making larger contributions to the environmental problems under discussion should bear larger burdens with respect to addressing the problems.

There is, of course, one final class of measures we could consider. Rather than just trying to incentivize people to procreate less, we could simply mandate it – that is, impose strict legal penalties on people who have more than a certain number of children. China's one-child policy is the clearest recent example of such a scheme. Policies of this sort do not tend to be popular, but that is not in itself a reason to reject them. We should consider if there is a moral case to be made for these policies in light of the long-term impacts of our growing population size.

Severely coercive measures

Everyone will readily agree that coercive population policies involve a serious infringement on a person's procreative autonomy. The main justification for implementing them appeals to their long-term benefits. Coercive policies can be effective in limiting population growth: the one-child policy in China prevented at least 500 million births between 1970 and 2000 (Lee and Liang 2006). That is a sizeable benefit, but as I have mentioned previously, these policies have often involved severe human rights violations in the form of forced abortions or sterilizations. They are also often associated with sex selection: in cultures where men are valued more than women, they create an incentive to abort fetuses identified as female and have another child in the hope that it is a boy.[7]

Now some regard these historical injustices as so serious that coercive population policies could never be deemed permissible, but that conclusion does not follow. If circumstances are dire enough, otherwise impermissible actions can become permissible. Killing an innocent person is one of the worst crimes one can commit, but if killing one innocent person is required to save the lives of ten other innocent people, then such a killing may well be morally permissible. In this manner, few (if any) broad moral principles are absolute. So while we recognize that the human rights violations that took place as a result of coercive population policies in the past were heinous and deplorable, there are at least *possible* circumstances in which the risk of these abuses would be worth taking.

We should also note that a strict and coercive population policy could take many forms. Sarah Conly (2015, 2016) provides the most in-depth recent defense of coercive population policies, but her proposal involves different enforcement mechanisms than mandatory abortion or sterilization. While she thinks that a one-child policy is permissible when the harms caused by overpopulation are severe enough, she believes it should be enforced through economic penalties and not by bodily invasions (Conly 2016, ch. 4). Even so, given the other options that have been discussed so far, any policy that imposes serious financial penalties for procreating would have to be a last resort – a final measure implemented solely for the sake of avoiding catastrophe after we have exhausted our other options. Even then, I believe that we would have compelling reasons not to consider such a policy.

For one, it is not clear that a one-child policy, even if enacted globally, would be the best means of lowering fertility rates. Citing data from Bradshaw and Brook (2014), Conly (2016) notes that dropping the fertility rate to one per woman by 2045 through full or nearly full compliance to a global one-child policy would shrink the population to 3.45 billion by 2100 (p. 219). That would indeed be a drastic reduction in human population, but a global one-child policy would never decrease the fertility rate to that extent. As Travis Rieder (2016) mentions, the one-child policy in China, which was more extreme than the kind of policy that Conly would endorse, only lowered fertility rates to an average of 1.6 children per couple (p. 33). Moreover, the fact that many European countries already have fertility rates comparable to this figure indicates that other strategies for reducing fertility rates can be just as effective.

A global one-child policy, or something approximating it, faces other objections as well, even if coercive bodily invasions are completely avoided. One of these is that applying a one-child policy to everyone would be unfair. Conly (2016) stresses the importance of equality in the context of exercising our rights, and on this basis, she argues that the constraints on procreation "must apply equally to everyone – not more children for some and fewer for others" (Conly 2016, p. 92). The problem with this position is that the constraints on procreation are being proposed in response to a problem where the contributions to it are *not* equal. Restricting everyone's procreation equally suggests that everyone has made a roughly equal contribution to the problem, and that is just not the case.[8] A fairer way to impose constraints on procreation will impose harsher constraints on those who have made larger contributions to the problem and lighter constraints on those who have made smaller contributions.[9]

A further objection is that a global one-child policy would be racist in its practical application.[10] The countries with the highest fertility rates in the world tend to be in Africa whereas the countries with the lowest fertility rates tend to be in Europe and North America. Thus, it would generally be far harder for people living in Africa to comply with a one-child policy than it would be for those in Europe and North America to do so. Perhaps more troubling than the racial inequality in the implementation of this policy would be the message that it might carry – namely, that African populations are in greater need of being controlled or regulated than those in whiter nations. Even if this result is unintentional, its moral significance cannot be ignored.

There is also a compelling practical reason not to pursue a one-child policy or anything resembling it: doing so will almost surely be counterproductive to the general goal of reducing population. Coercive population policies have been widely condemned, and the repulsion people feel toward them has played a considerable role in silencing discussion about population. In democratic societies existing at this stage of the twenty-first century, coercive population policies are not viable because they will never garner the necessary support among citizens. The only likely result of pushing for them is that people will become more reluctant to discuss population at all, which would reduce the likelihood of getting people to seriously consider other measures that could aid in reducing fertility rates. In this manner, advocating for severely coercive population policies with the aim of reducing population is self-defeating.

For all the reasons listed in this section, one-child policies and similar overtly coercive measures of regulating fertility cannot be part of our response to rising population. We will have to use other strategies to decelerate population growth.

Moral tragedies and difficult decisions

Even if we avoid coercive measures to reduce population growth, the transition toward population reduction still presents moral challenges. One of the unfortunate realities of our predicament is that it is probably impossible to respond to the population problem in a way that avoids all unjust outcomes. We know that failing to act will lead to substantial harm to future people – a great injustice. But there are also effects of pursuing the path to population reduction that may also result in significant injustice. Call these kinds of scenarios *moral tragedies* – situations where we cannot perform any action that avoids all unjust outcomes. I have already highlighted the wrongs that will befall future people if no efforts are made to constrain our burgeoning population size, but so far, the only notable cost to presently existing people that I have mentioned is a reduction in their procreative autonomy – a cost that might not even fully materialize if the autonomy-enhancing measures discussed earlier are successful. We should pause to consider some of the other costs to those in the present.

One short-term concern about decreasing population is that there will be too few young members of the population relative to the number of elderly people (Last 2013, ch. 5). One consequence of having a smaller working population is that the tax base declines, decreasing government revenue. Another is that there is an increased demand for medical care, which requires the government to spend more on health coverage. This combination of effects creates a significant dilemma: either the young, working members of society must bear a greater burden to support the elderly, or medical care to the elderly must be more strictly rationed. Independent of any connection to population reduction, some have argued that we ought to ration life-extending health care on the grounds that medical resources are limited and that keeping the very old alive for a bit longer through expensive procedures is an inappropriate use of limited resources (Callahan 1995, 2012).[11] The need to reduce population would seem to make the case for such rationing even stronger, but of course, doing so means that some older members of society will not receive treatments that could extend their lives. We may also confront more direct conflicts between the pursuit of population reduction and the maintenance of adequate medical care, such as if we must choose whether to fund family planning or health care (Mosher 2008, ch. 6).

A related concern is that a reduction in population growth will stifle or deter economic growth. Lower population growth, the thought goes, will lead to fewer consumers and fewer workers, and the result will be decreased economic activity. This line of reasoning is somewhat intuitive, but the empirical reality of the relationship between population and economic growth is not as straightforward as it suggests (Peterson 2017). Economic growth is affected by many factors beyond the fertility rate, and some recent evidence suggests that lower fertility is compatible with both lowering carbon emissions *and* increasing income per capita (Casey and Galor 2017). Additionally, a goal of endless economic growth is incompatible with existing in a world of finite resources. At some point, we must transition to an economic model grounded in the sustainable use of resources rather than one based on ceaseless expansion and consumption (Cafaro 2015, pp. 170–171;

Daly 1991). Given the environmental degradation that is taking place around the world, the evidence is mounting that we should begin that transition sooner rather than later.

Population policies also raise significant concerns about equality. Women bear a much larger role in reproduction than men, and so these policies may have a disproportionate impact on them. Rebecca Kukla (2016) argues that these policies "will likely enhance an already problematic pattern of gender inequality, and intensify our interventionism and moralism then it comes to women's bodies and reproductive practices" (p. 876). The main fear is that women will be subject to substantial pressure from others regarding their reproductive decisions and that, particularly in societies where their reproductive freedom is already compromised, their autonomy will be undermined. These concerns will be most pronounced if we are considering incentivization schemes, and some of these concerns can be mitigated by avoiding certain types of incentives. To offer one illustration, Hickey, Rieder, and Earl (2016) discuss paying women to attend family planning classes or visit a gynecologist (p. 867). Such incentives might be effective, but they seem to target women exclusively, suggesting that it is primarily a woman's responsibility to limit her fertility. Incentives should strive to be gender neutral. Even so, given the prevailing view that women are the ones who are primarily responsible for their reproductive activities, it is probably naïve to think that these types of population policies could *completely* avoid having a disproportionate impact on women.

Another concern about inequality stems from the disproportionate impact certain policies may have on the poor. Those who occupy lower socio-economic classes will be heavily incentivized to have smaller families to procure financial benefits (or avoid penalties), so the fear is that large families may become common only among the very wealthy. Wealthier people would be more easily able to cope with foregone financial benefits or purchase entitlements for additional children. In this manner, family size might become associated with social class. More worryingly, since the poor are disproportionately likely to be people of color, these policies could "end up enacting a kind of indirect eugenics" (Kukla 2016, p. 877). The extent to which this outcome would actually manifest is debatable – in practice, more affluent people usually choose to have *fewer* children than other people (Bognar 2019, p. 326). Even so, this implication does give us further reason to hesitate in our adoption of incentivization schemes.

It is clear that taking serious measures to reduce population growth will impose some costs on present people and that not doing so will impose some significant costs on future people. So how do we decide what to do in this morally tragic situation? It is not possible to do justice to all parties involved or protect all parties from harm, so the best we can do is to minimize the injustice that occurs and the harm that is suffered.

One way to pursue this strategy is adopting a consequentialism of rights, a strategy discussed by Darrel Moellendorf (2014) in the context of climate change mitigation. He recognizes the possibility that some people who will not have their human rights violated under business-as-usual scenarios will have their human rights violated if we undertake mitigation measures (Moellendorf 2014, pp. 231–232). If this picture is accurate, then one may wonder how a rights-based approach could favor a policy of mitigation rather than business-as-usual. After all, rights are being violated in both scenarios. Moellendorf (2014) entertains the possibility that this problem might be resolved by pursuing "the course of action that is likely to lead to maximal satisfaction of rights" (p. 232).[12]

One complication to this consequentialism of rights is that some rights violations are worse than others. Violating someone's right to life is a more serious moral wrong than violating someone's right to bodily autonomy, though both rights are significant. The rights that will be violated as a result from unimpeded environmental degradation will be among the most severe (e.g., the right to life, the right to health, the right to physical security). These rights violations could be experienced by hundreds of millions of people this century. Given the staggering numbers and the severity of the rights violations under discussion, we should prioritize reducing population to avoid these rights violations and accept that some rights violations – perhaps in the form of inequality or unintentional coercion – will be experienced by present people as a result, despite our best efforts to avoid these outcomes. These results are regrettable, but it would be morally worse for us to *not* take these measures to respond to population growth.

This resolution may sound dissatisfying. It would be preferable to arrive at a solution in which all parties can be treated fairly and protected from harm. But our circumstances have made such a solution impossible, and we do ourselves no favors by denying this fact. Moreover, as the survey in this chapter shows, we still have some fairly good non-coercive options that we can employ to reduce population growth. But the longer we wait to act, the harder it will be to make the reductions in our collective ecological footprint in time to avert serious harms. If things get significantly worse, then the need to seriously consider more coercive measures could arise. This is just a further reason why the better option is to pursue population reduction *now*.

What should be done

We have a wide array of policy measures we could pursue in response to population growth. Among the options available, autonomy-enhancing measures are the least controversial and most beneficial. We should make every effort to increase access to contraception and family planning services, improve sex education, and promote gender equality: these measures would lead to lower fertility rates while also increasing procreative autonomy. Preference adjusting interventions through media campaigns, so long as they are not done deceptively, should also be undertaken to counteract pronatalist values and encourage greater reflection on procreative choices.

Incentive-based schemes, even if they are effective, encounter a number of obstacles – both moral and practical. Since there is so much variance in how these schemes could be designed, it would be too hasty to rule out all of them, but we would need to exercise a great deal of caution in how they were implemented. Proposals for incentive-based schemes would have to be evaluated on a case-by-case basis, and I suspect few of them could be designed in ways that sufficiently minimized the injustices that could result from them. Overall, I favor a cautious approach toward incentivization: we should exhaust our other permissible options first, and even after that, we should subject any proposed incentive-based scheme to substantial scrutiny before we consider implementing it.

Finally, we should not implement outright coercive options that involve strict, government-mandated limits on how many children people can have. These policies run a high risk of causing severe harm and injustice, and they are not likely to be more effective than the myriad of other options that are available to us. Moreover, presenting serious proposals to implement these policies would be counterproductive: it would likely push people away from the subject of population growth in the same way it did in the past.[13]

Notes

1. As I will discuss later, however, we may face circumstances where we must seriously consider rationing health care to the elderly.
2. As revealed by some of the information about the United States, there is room for improvement with respect to sex education in some developed nations.
3. In cases where pregnancy results from rape, the woman would have plenty of time to determine that she was pregnant, deliberate about whether to carry the fetus to term, and then get an abortion (if she chooses) within the first trimester. Thus, allowing abortions in response to involuntary pregnancy may not require any special provision that extends beyond the first trimester.
4. Additionally, Lovering (2005) questions whether a fetus really has a future prior to being conscious. He reasons that only psychologically continuous entities appear to have futures in a morally relevant sense, and consciousness is a prerequisite for this kind of psychological continuity. If Lovering's view is correct, then Marquis' position may not turn out to be much different than the view I have sketched above – where abortions are permissible in the first trimester but often prohibited afterward. Abortions would be permissible during the portion of the pregnancy where the fetus lacks consciousness and impermissible thereafter (since the fetus would then have a future that warrants protection).
5. Ideally, we would eliminate the possibility of injustice altogether, but very few social policies can be constructed in ways such that they *never* lead to injustice.
6. It is still important to encourage lower fertility in the developing world so that their ecological footprints do not balloon dramatically as they develop, but this can be done without ignoring the disparity in per capita ecological footprint between these countries and the world's wealthiest nations.
7. Conly (2016) points out that the main cause of sex selection is the prevalence of sexist attitudes in the background culture of these societies rather than coercive population policies as such (pp. 193–204). Even so, in practice, the fact that a

strict limit on the number of children a couple can have could exacerbate gender inequality remains a strong reason to oppose the implementation of coercive population policies.
8 Chen (2017) also suggests that Conly's one-child policy may be unfair if it is applied globally (p. 453).
9 In this manner, our population policies should strive to be consistent with the Polluter Pays principle – the notion that those who contribute to the problem should bear the burdens of solving the problem or compensating the victims, at least in cases where the pollution is not caused by excusable ignorance. For a critical appraisal of the Polluter Pays principle in the case of Climate Change, see Caney (2010).
10 I have raised this same concern elsewhere. See Hedberg (2017b).
11 For a recent overview of the issues involved in rationing health care, see Morreim et al. (2014).
12 This maximizing approach to human rights does run the risk of not according strong enough protections to the rights of minorities: the rights of the majority would appear to always trump the rights of minorities in rights conflicts between these groups. This may justify sometimes giving the rights of minorities disproportionate weight in the calculation. Fortunately, this consideration is not relevant to the case we are addressing because the interests of minority groups will be jeopardized in both of the scenarios we are considering – whether we take deliberate action to reduce population growth or whether we avoid doing so.
13 Significant portions of this chapter are derived from chapter 6 of my doctoral dissertation. See Hedberg (2017a).

References

Advocates for Youth. 2007. *The Truth about Abstinence-Only Programs.* www.advocatesforyouth.org/wp-content/uploads/storage//advfy/documents/fsabstinenceonly.pdf. Accessed December 8, 2019.

Alvarado, Susana, and Jaqueline Echegaray. 2010. "Going to Extremes: Population Politics and Reproductive Rights in Peru." In *A Pivotal Moment: Population, Justice and the Environmental Challenge*, 2nd ed., edited by Laurie Mazur, 292–299. Washington, D.C.: Island Press.

Bognar, Greg. 2019. "Overpopulation and Procreative Liberty." *Ethics, Policy & Environment* 22, no. 3: 319–330.

Bongaarts, John, and Steven Sinding. 2011. "Population Policy in the Developing World." *Science* 333, no. 6042: 574–576.

Bradshaw, Corey, and Barry Brook. 2014. "Human Population Reduction is Not a Quick Fix for Environmental Problems." *Proceedings of the National Academy of Sciences* 111, no. 46: 16610–16615.

Breuner, Cora, and Gerri Mattson. 2016. "Sexuality Education for Children and Adolescents." *Pediatrics* 138, no. 2: e20161348. DOI: 10.1542/peds.2016-1348.

Cafaro, Phil. 2012. "Climate Ethics and Population Policy." *Wiley Interdisciplinary Reviews: Climate Change* 3, no. 1: 65–81.

Cafaro, Philip. 2015. *How Many Is Too Many? The Progressive Argument for Reducing Immigration into the United States.* Chicago: University of Chicago Press.

Callahan, Daniel. 1995. *Setting Limits: Medical Goals in an Aging Society.* Washington D.C.: Georgetown University Press.

Callahan, Daniel. 2012. "Must We Ration Health Care to the Elderly?" *Law, Medicine & Ethics* 40, no. 1: 10–16.

Caney, Simon. 2010. "Climate Change and the Duties of the Advantaged." *Critical Review of International Social and Political Philosophy* 13, no. 1: 203–228.

Casey, Gregory, and Oded Galor. 2017. "Is Faster Economic Growth Compatible with Reductions in Carbon Emissions? The Role of Diminished Population Growth." *Environmental Research Letters* 12, no. 1: 014003. DOI: 10.1088/1748-9326/12/1/014003

Chen, Jason. 2017. Review of *One Child: Do We Have a Right to Have More? Journal of Applied Philosophy* 34, no. 3: 452–453.

CIA (Central Intelligence Agency). 2019. "Country Comparison: Total Fertility Rate." *The World Factbook*. www.cia.gov/library/publications/the-world-factbook/rankorder/2127rank.html. Accessed December 7, 2019.

Conly, Sarah. 2015. "Here's Why China's One-Child Policy Was a Good Thing." *Boston Globe*. www.bostonglobe.com/opinion/2015/10/31/here-why-china-one-child-policy-was-good-thing/GY4XiQLeYfAZ8e8Y7yFycI/story.html. Accessed December 8, 2019.

Conly, Sarah. 2016. *One Child: Do We Have a Right to Have More?* Oxford: Oxford University Press.

Cripps, Elizabeth. 2016. "Population and Environment: The Impossible, the Impermissible, and the Imperative." In *The Oxford Handbook of Environmental Ethics*, eds. Stephen Gardiner and Allen Thompson, 380–390. Oxford: Oxford University Press.

Crist, Eileen. 2019. *Abundant Earth: Toward an Ecological Civilization*. Chicago: University of Chicago Press.

Daly, Herman. 1991. *Steady State Economics*, 2nd ed. Washington, D.C.: Island Press.

de la Croix, David, and Axel Gosseries. 2009. "Population Policy Through Tradable Procreation Entitlements." *International Economic Review* 50, no. 2: 507–542.

Ford, Liz. 2016. "Rise In Use of Contraception Offers Hope for Containing Global Population." *Guardian*. www.theguardian.com/global-development/2016/mar/08/rise-use-contraception-global-population-growth-family-planning. Accessed December 8, 2019.

Global Footprint Network. 2019. "Ecological Footprint Per Capita." https://data.footprintnetwork.org/#/compareCountries?cn=all&type=EFCpc&yr=2016. Accessed December 8, 2019.

Grossu, Arina. 2017. *What Science Reveals about Fetal Pain*. http://downloads.frc.org/EF/EF15A104.pdf. Accessed December 8, 2019.

Guttmacher Institute. 2019. *Sex and HIV Education*. www.guttmacher.org/state-policy/explore/sex-and-hiv-education. Accessed December 8, 2019.

Hedberg, Trevor. 2017a. "Population, Consumption, and Procreation: Ethical Implications for Humanity's Future." Ph.D. dissertation, Department of Philosophy, University of Tennessee.

Hedberg, Trevor. 2017b. Review of *One Child: Do We Have a Right to Have More? Philosophy East and West* 67, no. 3: 934–938.

Heil, Sarah, Diann Gaalema, and Evan Herrmann. 2012. "Incentives to Promote Family Planning." *Preventative Medicine* 55, no. Suppl: S106–S112.

Hesketh, Therese, and Wei Xing Zhu. 1997. "The One Child Family Policy: The Good, the Bad, and the Ugly." *British Medical Journal* 314, no. 7095: 1685–1687.

Hickey, Colin, Travis Rieder, and Jake Earl. 2016. "Population Engineering and the Fight against Climate Change." *Social Theory and Practice* 42, no. 4: 845–870.

Johnson, Steve. 2016. "Africa's Life Expectancy Jumps Dramatically." *Financial Times*. www.ft.com/content/38c2ad3e-0874-11e6-b6d3-746f8e9cdd33. Accessed December 9, 2019.

Kukla, Rebecca. 2016. "Whose Job Is It to Fight Climate Change? A Response to Hickey, Rieder, and Earl." *Social Theory and Practice* 42, no. 4: 871–878.

Kweifio-Okai, Carla, and Josh Holder. 2016. "Over-populated or Under-developed? The Real Story of Population Growth." *Guardian*. www.theguardian.com/global-development/datablog/2016/jun/28/over-populated-or-under-developed-real-story-population-growth. Accessed December 8, 2019.

Last, Jonathan. 2013. *What to Expect When No One's Expecting: America's Coming Demographic Disaster*. New York: Encounter Books.

Lee, Che-Fu, and Quisheng Liang. 2006. "Fertility, Family Planning, and Population Policy in China." In *Fertility, Family Planning, and Population Policy in China*, edited by Dudley Poston, Jr., Che-Fu Lee, Chiung-Fang Chang, Sherry McKibben, and Carol Walther, 159–171. London: Routledge.

Lee, Susan, Henry Ralston, Eleanor Drey, John Partridge, and Mark Rosen. 2005. "Fetal Pain: A Systematic Multidisciplinary Review of the Evidence." *Journal of the American Medical Association* 294, no. 8: 947–954.

Lovering, Robert. 2005. "Does a Normal Fetus Really Have a Future of Value? A Reply to Marquis." *Bioethics* 19, no. 2: 131–145.

Marquis, Don. 1989. "Why Abortion Is Immoral." *Journal of Philosophy* 86, no. 4: 183–202.

Mazur, Laurie. 2010. "Introduction." In *A Pivotal Moment: Population, Justice and the Environmental Challenge*, 2nd ed., edited by Laurie Mazur, 1–23. Washington, D.C.: Island Press.

Moellendorf, Darrel. 2014. *The Moral Challenge of Dangerous Climate Change: Values, Poverty, and Policy*. New York: Cambridge University Press.

Morreim, Haavi, Ryan Antiel, David Zacharias, and Daniel Hall. 2014. "Should Age Be a Basic for Rationing Health Care?" *Virtual Mentor* 16, no. 5: 339–347.

Mosher, Steven. 2008. *Population Control: Real Costs, Illusory Benefits*. New Brunswick, NJ: Transaction Publishers.

Peterson, E. Wesley. 2017. "The Role of Population in Economic Growth." *SAGE Open* 7, no. 4: 215824401773609. DOI: 10.1177/2158244017736094.

Population Reference Bureau. 2016. "2016 World Population Data Sheet." www.prb.org/pdf16/prb-wpds2016-web-2016.pdf. Accessed December 8, 2019.

Pritchett, Lant. 1994. "Desired Fertility and the Impact of Population Policies." *Population and Development Review* 20, no. 1: 1–55.

Repetto, Robert. 1968. "India: A Case Study of the Madras Vasectomy Program." *Studies in Family Planning* 31: 8–16.

Rieder, Travis. 2016. "Review: Sarah Conly, *One Child: Do We Have a Right to Have More?*" *Kennedy Institute of Ethics Journal* 26, no. 2: 29–34.

Rogers, Everett, Peter Vaughan, Ramadhan Swalehe, Nagesh Rao, Peer Svenkerud, and Suruchi Sood. 1999. "Effects of an Entertainment-education Radio Soap Opera on Family Planning Behavior in Tanzania." *Studies in Family Planning* 30, no. 3: 193–211.

Roudi-Fahimi, Farzaneh, and Mary Mederios Kent. 2007. "Challenges and Opportunities – The Population of the Middle East and North Africa." *Population Bulletin* 62, no. 2: 1–19. Washington, D.C.: Population Reference Bureau. www.prb.org/pdf07/62.2MENA.pdf. Accessed December 8, 2019.

Ryerson, William. 2010. "Population: The Multiplier of Everything Else." In *The Post Carbon Reader: Managing the 21st Century's Sustainability Crises*, eds. Richard Heinberg and Daniel Lerch, 153–174. Healdsburg, CA: Watershed Media.

Ryerson, William. 2012. "How Do We Solve the Population Problem?" In *Life on the Brink: Philosophers Confront Population*, eds. Phil Cafaro and Eileen Crist, 240–254. Athens, GA: University of Georgia Press.

Sedgh, Gilda, Susheela Singh, and Rubina Hussain. 2014. "Intended and Unintended Pregnancies Worldwide in 2012 and Recent Trends." *Studies in Family Planning* 45, no. 3: 301–314.

Singhal, Arvind, and Everett Rogers. 1989. *India's Information Revolution*. New Delhi: Sage.

Sinnott-Armstrong, Walter. 1999. "You Can't Lose What You Ain't Never Had: A Reply to Marquis on Abortion." *Philosophical Studies* 96, no. 1: 59–72.

Sinnott-Armstrong, Walter. 2005. "It's Not My Fault: Global Warming and Individual Moral Obligations." In *Perspectives on Climate Change: Science, Politics, Ethics*, eds. Walter Sinnott-Armstrong and Richard B. Howarth, pp. 285–307. Amsterdam: Elsevier.

Stranger-Hall, Kathrin, and David Hall. 2011. "Abstinence-Only Education and Teen Pregnancy Rates: Why We Need Comprehensive Sex Education in the United States." *PLoS One* 6, no. 10: e24658. DOI: 10.1371/journal.pone.0024658.

Sumner, L.W. 1981. *Abortion and Moral Theory*. Princeton, NJ: Princeton University Press.

Tawia, Susan. 1992. "When Is the Capacity for Sentience Acquired During Fetal Development?" *Journal of Maternal-Fetal Medicine* 1, no. 3: 153–165.

Thomas, Neil. 1995. "The Ethics of Population Control in Rural China." *Population, Space, and Place* 1, no. 1: 3–18.

Thomson, Judith Jarvis. 1971. "A Defense of Abortion." *Philosophy and Public Affairs* 1, no. 1: 47–66.

UN Department of Economic and Social Affairs, Population Division. 2015. *Trends in Contraceptive Use 2015*. www.un.org/en/development/desa/population/publications/pdf/family/trendsContraceptiveUse2015Report.pdf. Accessed December 8, 2019.

United Nations Department of Economic and Social Affairs, Population Division. 2019. *World Population Prospects 2019: Volume I: Comprehensive Tables*. https://population.un.org/wpp/Publications/Files/WPP2019_Volume-I_Comprehensive-Tables.pdf. Accessed December 5, 2019.

Warren, Mary Anne. 1973. "On the Moral and Legal Status of Abortion." *The Monist* 57, no 1: 43–61.

6 Individual procreative obligations

In the previous chapter, I examined the policies that we could implement to reduce population growth at the global level and hasten population reduction in nations that are already below replacement-level fertility. However, those policy measures and the collective obligations that give rise to them do not exhaust the moral landscape with respect to environmentally responsible procreation. There remain significant moral questions for individuals as well. Specifically, for individuals who are considering whether to have children, what moral principles should guide their decision-making? Are they obligated to limit their family size? If so, what is the maximum number of biological children they ought to have?

In this chapter, I turn my attention to these questions tied to individual procreative obligations. I argue that those who have control over their reproductive choices should limit their biological offspring to one child per person (two children per couple).[1] Additionally, while I believe this argument generalizes to some extent, it is primarily aimed at people in the world's wealthiest nations because residents of these countries have the largest per capita ecological footprints and because they are likely to enjoy a much greater degree of reproductive freedom than those living in other parts of the world. I first argue that we have a moral duty to limit our individual environmental impacts and draw in part on discussions about the duty to limit one's individual contribution to climate change to present this argument. I then explain why this obligation entails a duty to limit family size. Lastly, I specify the limits of permissible procreation and answer some objections to my position.

The individual obligation to limit one's environmental impact

There are many potential routes for arguing that we have a moral obligation to reduce our individual ecological footprints. Due to the robust literature on climate change, many of these arguments have been discussed in depth with regard to the narrower obligation to reduce our individual carbon footprints. While the environmental impact of procreation is not limited to increased carbon emissions, focusing on this dimension of the problem provides a useful starting point because most of the arguments (if successful) would generalize from a duty to reduce one's carbon footprint to a broader duty to reduce one's ecological footprint. So I will start with a quick overview of some arguments that support obligations for individuals to reduce their carbon footprints.

Avoiding harm to future people

As I mentioned in Chapter 3, one of the most basic moral imperatives is the duty to avoid unnecessary harm. This basic moral principle can be used to make an argument against unnecessary carbons emissions if it can be established that they cause harm. John Nolt (2011) adopts this strategy and argues that the lifetime greenhouse gas emissions of the average American are responsible for the severe suffering or death or 1–2 future people. He arrives at this conclusion by calculating the total lifetime emissions of a typical American born in 1965 and then estimating the total amount of greenhouse gas emissions that they are likely to emit. According to his math, the average American contributes about one two-billionth of global greenhouse gas emissions. That's a miniscule proportion, but because the effects of climate change will be so long-lasting, billions of present and future people may be impacted. If two billion people suffer severely or die from climate change and a person's moral responsibility for that harm is proportional to their contribution to the problem, then the average American is morally responsible for causing a future person to suffer severely or die.[2]

Similarly, John Broome (2012) appeals to a principle of non-harm to defend an individual obligation to have zero GHG emissions, whether that is achieved by not emitting any GHGs at all or by offsetting one's emissions (pp. 73–96). On the basis of estimates from the World Health Organization (2009) about the deaths caused by global warming, he states that the lifetime emissions of a person from a wealthy nation will "wipe out more than 6 months of healthy human life" (Broome 2012, p. 74). Moreover, even though Broome acknowledges that daily emissions have imperceptible consequences, he argues that tiny imperceptible harms can nonetheless add up to significant harm. Since individuals have a moral duty to avoid unnecessary harm, Broome regards them as having an obligation to be net-zero carbon emitters.

Some regard this harm-based argument as convincing, but many philosophers are not so confident it succeeds. While the imperative not to cause harm is a fundamental moral principle, it is not obvious to everyone that an individual person's GHG emissions cause harm in the way this argument supposes. What happens to our particular emissions can vary rather dramatically. Some of them may stay in the atmosphere, but others may become stored in a natural carbon sink or dissolve in the ocean. Those GHG emissions that stay in the atmosphere could lead to a harmful effect, such as raising the sea level just enough to render a small island uninhabitable or intensifying the winds in a hurricane just enough to destroy property that would have otherwise been spared. But they might also have benign effects such as causing a heat wave in an uninhabited part of the world or increasing temperature in the Siberian summer just enough to slightly improve crop yields. Thus, for a particular act of emitting GHGs, it is uncertain whether these emissions stay in the atmosphere and whether they emissions are causally connected to the harmful effects of climate change. Additionally, my individual emissions are harmless in isolation: it is only when they are joined with the emissions of millions of other people that any negative effects are generated. This combination of factors has led many philosophers to question the claim that an individual's GHG emissions cause harm (e.g., Jamieson 2014a, 2014b, pp. 144–169; Kawall 2011; Rieder 2016, pp. 13–27; Sinnott-Armstrong 2005).

Do these considerations show that a harm-based duty to reduce GHG emissions is unfounded? Not necessarily. Defenders of this harm-based duty have made some creative attempts to show that the connection between individual action and harm is strong enough to ground a moral obligation (e.g., Almassi 2012; Broome 2012, pp. 75–78; Cripps 2011; Fragnière 2018, pp. 654–656; Lichtenberg 2010; Nolt 2013a; Schwenkenbecher 2014). Nevertheless, defending a duty to reduce GHG emissions along these lines is bound to be difficult and yield uncertain results. The core problem with this line of argument – namely, the nebulous causal relationship between a particular act of emitting and an actual harmful effect experienced by someone else – is difficult to overcome.[3] A similar problem will arise if we attempt to broaden this obligation to encompass a reduction of our ecological footprints. The causal story of how our individual acts of consumption impact the world is similarly complex. If an individual obligation to reduce our individual ecological footprints is to be defended, it would be preferable to appeal to an argument that does not rely on a contentious account of harm. I believe there two promising arguments that are up to the task.

Not contributing to injustice

Suppose that your individual GHG emissions do not cause harm. Does this excuse you from reducing your emissions? I think not. One reason is that your actions – even if they are not causally connected to harmful outcomes in the way required for us to say they *cause* harm – are nonetheless contributing to the causal process that leads to the harm. Doing so is often wrong. Consider Americans living in the United States before the Civil War who understood the moral horrors of slavery but continued to buy cheap goods that resulted from slave labor. Their economic contributions may well have made no difference to the demand for slave labor and may not have caused any harm to actual slaves. Would we be willing to say these Americans are doing nothing wrong through purchasing slave-produced goods? Or imagine a scientist who is employed by a corrupt organization or governmental regime that uses its power to harm innocent people. The scientist does not produce weapons or other items that directly perpetrate these harms, but the knowledge this person produces does play some role in the organization's functioning, even if it does not ultimately cause harm to befall someone. Would we be willing to say that scientist is doing nothing wrong by working for this organization?[4]

Travis Rieder suggests that these actions are morally wrong but that their wrongness cannot be explained by an appeal to causing harm. In these cases, there exists a complex system that causes harm. The individuals we have considered are participating in this system but not necessarily causing harm themselves. He suggests that the wrongdoing in these cases originates from violating a duty not to contribute to massive, systematic harms – "a duty not to inject oneself as an active contributor into the large, causally complex machine that is doing the harm" (Rieder 2016, p. 29). This duty captures an important conviction that many of us have. We think that it is morally wrong to play a role – even a causally impotent one – in an ongoing injustice when we can refuse to participate at little cost to ourselves.

Climate change is one example of a massive, systematic harm. The phenomenon causes harm to present and future people, but its causal mechanism is complicated in ways that make individual contributions seem insignificant. Even so, if we want to uphold our duty not to contribute to massive, systematic harms, then we should try to limit our individual carbon footprints. Doing so may not make a difference, but it will at least serve as evidence that we are trying to minimize the role we play in the process that causes harm. That is probably the best we can do since few of us will be able to eliminate carbon emissions from our lives completely without leaving ourselves destitute.

We can reason the same way about environmental degradation more generally – most of these processes are massive, systematic harms. They result from complex processes tied to transportation, agriculture, waste management, energy use, and other aspects of our economic and social infrastructure. In our efforts to avoid contributing to these harms, we should try to limit our individual contributions to environmental degradation. No decision is more impactful on the size of our individual ecological footprints than the number of children we choose to have. Our children will grow up in the same society that we inhabit and are likely to adopt many of the same behaviors that we do. So if I live in a developed country, every new child I have will leave a massive ecological footprint.

Take the carbon footprint tied to procreation as an example. In the United States, driving a vehicle that gets 30 mpg rather than 20 mpg for your entire life will save you 148 metric tons of CO_2. That's no small sum, but it's less than 1.6 percent of the 9441 metric tons of CO_2 that you would save by having one fewer child (Murtaugh and Schlax 2009). In their assessment of which lifestyle changes had the biggest impact on a person's carbon emissions, Wynes and Nicholas (2017) concluded that for a person living in the developed world, not having an additional child would (on average) save

58.6 tonnes of CO_2 equivalent annually. The next highest impact action in their survey was living car free – which would save an average of 2.4 tonnes of CO_2 equivalent annually. If this calculation is accurate, then choosing to have one fewer child is 24 times more effective at reducing one's carbon footprint than living without a car! Perhaps more incredibly, having one fewer child makes a six times larger difference than the sum of all 11 of the other individual actions they examined.

Given that our procreative activities involve a much larger contribution to environmental degradation than any of our other individual activities, the duty not to contribute to massive systemic harms surely entails that we should restrict our procreation to some extent. Rieder (2016) reaches precisely this conclusion and suggests that those who procreate must justify their choice to procreate in light of powerful moral reasons against procreation.[5] While he does not say with certainty that the limit of permissible procreation is two children, he has "real trouble coming up with plausible justifications" for having three or more children (Rieder 2016, p. 64).

I view the duty not to contribute to massive, systematic harms to be a plausible account of how we should act in the presence of complex collective action problems that cause great harm. However, readers who are inclined toward consequentialist reasoning may resist. If an action really does not make a difference to the overall outcome, then they may reason that they have no genuine moral reason to refrain from doing it. To appease these readers, I will offer one more argument in favor of a duty to restrict our procreative activities in light of overpopulation.

Maintaining personal integrity

In previous work (Hedberg 2018a), I have appealed to the virtue of integrity to defend a moral duty to reduce our individual carbon footprints. At its core, the virtue of integrity involves having a consistent set of beliefs and values and acting in accordance with those values even in spite of temptations to compromise them.[6] To express the idea as an aphorism, a person of integrity is someone who practices what they preach. The antithesis of integrity is hypocrisy, which manifests when a person acts in ways that are inconsistent with the values they hold (or, at least, the values they claim to hold). Hypocrites are sometimes blameworthy for the ways in which they deceive other people, but even in cases where their hypocrisy does not cause harm to others, it is still routinely viewed as a failing of their character.

We value integrity for several reasons. First, it accords with basic human psychology. Normally, people are made uncomfortable when they recognize that their values are inconsistent or that they are not behaving in ways that align with their values.[7] One reason to live with integrity is to avoid this unpleasant feeling. People are often happier when the different spheres they inhabit are unified by a common set of values, so if one of our goals is to live

Individual procreative obligations 89

a flourishing life, maintaining integrity will probably be part of achieving that goal.

A second reason for valuing integrity is that it helps us explain what is praiseworthy about certain behaviors. Consider Thomas Hill's (1979) description of an old woman living in Nazi Germany:

> She lives on modest savings and offers no support to the Nazi regime either physically or morally. When the latest discriminatory laws against Jews are enforced, she is moved to protest. As a non-Jew she could have remained silent and thereby avoided much subsequent harassment. She is regarded as a silly eccentric and so cannot expect to make an impact on others, much less to stop the Nazi machinery. She still feels she should speak up, but she wonders why.
>
> (p. 84)[8]

This woman takes a stand against the Nazi regime even though it works against her self-interest and even though her protest is unlikely to contribute to solving the problem. These details suggest that her behavior is just irrational, but that cannot be the entire story. We often praise people who take symbolic stands to oppose practices that appear grossly unjust, even when their protests work to their individual disadvantage and do not make a difference to solving the problem. Why are such acts praiseworthy? One explanation is that individuals who engage in this behavior exemplify integrity. They have a deep moral conviction that a grave injustice is taking place, and their conviction is strong enough that they do not feel they can stay silent. This behavior is a paradigm example of acting in accordance with one's deeply held beliefs.

Finally, and perhaps most importantly, maintaining integrity is one way in which we convey the importance of our commitments to others. Marion Hourdequin (2010) highlights how this aspect of integrity can be especially important with respect to environmental issues:

> Interpersonally, integrity is a virtue from the perspective of intersubjective intelligibility and in affirming to others the authenticity of one's commitments. Where we see in others a lack of coherence between their political commitments and personal choices, we often wonder how to make sense of this apparent mismatch, and we may question the sincerity with which certain commitments are held. A politician's environmental commitments, as embodied in public pronouncements and legislative support, for example, may be called into question if he or she lives a lavish and environmentally damaging lifestyle.
>
> (p. 451)

In this manner, integrity is valuable in part because having it is a prerequisite for being able to enact meaningful political change. Acting with integrity can

90 *Ethics, policy, and obligations*

be a means of demonstrating the kind of change that one wants to see in the world and a way of demonstrating that one's convictions are sincere.[9] In this manner, the possession of integrity plays a role in getting others to take our causes seriously.

The social value of integrity is particularly noteworthy in the presence of large-scale collective action problems like climate change. Almost 30 years ago, when remarking on the tendency to approach climate change by focusing on probable outcomes, Dale Jamieson (1992) lamented that this way of engaging the issue tended to make people "cynical calculators" and to institutionalize hypocrisy (p. 150). Since our individual contributions to climate change are small and (seemingly) negligible, we can all reason that the effects of climate change seem fated to occur regardless of what we do as individuals. If that is the end of our deliberation, then we have no reason to change our individual behavior. If everyone follows that same pattern of reasoning, then the social change needed to respond to the issue on a collective level will never come to pass. If we are to avoid that result, we will need "people of integrity and character who act on the basis of principles and ideals" (Jamieson 1992, pp. 150–151). While the value of integrity cannot be reduced solely to its usefulness in solving climate change and other collective action problems, its role in this regard does give us a good reason to reject the claim that our individual actions make no difference.

So integrity is a character trait that we ought to cultivate in ourselves. But how do we get from that observation to a duty to reduce our individual carbon footprints? Here is an outline of the general argument:

1 Individuals should live with integrity.
2 Individuals should cooperate in working toward a collective political solution to climate change.
3 If individuals should live with integrity and cooperate in working toward a collective political solution to climate change, then they should reduce their personal greenhouse gas emissions.
4 Therefore, individuals should reduce their personal greenhouse gas emissions. [1–3][10]

I have already explained why integrity is valuable and a character trait we should cultivate in ourselves,[11] so the first premise is supported. Given the severity of climate change and the need for urgent action, the second premise should be an uncontroversial claim. Minimally, I suspect this duty would entail voting for politicians who support serious action on climate change (when such candidates are available) and being willing to endorse policies and other large-scale changes that would mitigate climate change, but for our purposes here, I will leave it open what else it may require.

The support for the third premise is derived from the notion that integrity requires coherence among our values and commitments. In this case, our commitments in the political sphere should align with our behavior in the personal sphere. So, if we adopt a political commitment to doing something about climate change, that commitment should also manifest in our individual behavior. Where we can make emissions reductions at little or no personal cost, we should make those emissions reductions. Having integrity means that we should not compromise these commitments unless we have good reasons to do so. This point is particularly salient with respect to carbon emissions because we are often tempted to increase our individual emissions in frivolous and unnecessary ways, usually by excessive energy consumption.

Now some have suggested to me that there would be nothing incoherent about adopting the political commitment to aid a collective response to climate change while also adopting a principle of only performing activities that would actually make a difference toward solving that problem.[12] If such a person reasons that their individual emissions make no difference to solving the problem, they may conclude that they have no moral duty to reduce their own individual emissions and should instead focus on actions that will be efficacious in enacting meaningful changes at the political level.

There are two responses to this line of argument. First, given the role that individual behavior plays in communicating our commitments to others, there is a significant risk that neglecting personal emissions reductions would hinder efforts to realize the desired political change. When people scrutinize the lifestyle of a high profile climate activist like Al Gore, they are trying to see if his own carbon footprint is a reflection of the political change he wants to see in the world. But the same concern arises at a smaller scale for ordinary people. It is much harder to convince others to support efforts to respond to climate change when our individual lifestyle sends the message that we are not actually all that concerned about climate change.

As a second response, I deny that this kind of consistency reflects the type of coherence that integrity demands. If we understand consistency among our values to be mere logical consistency, then indeed there is no conflict between adopting a commitment to supporting political action while also rejecting a commitment to do anything to change one's individual lifestyle. But integrity is not based on just maintaining a weak logical consistency among our values. We can highlight the problem with a few cases.

First, suppose a priest spends much of his time preaching about the importance and sanctity of monogamy. We later learn that he is in a polygamous relationship. Suppose he believes that the general public needs to think that monogamy should always be maintained but also believes that a commitment to monogamy is not important in his particular case. Second, imagine someone who advocates for donating large portions of our disposable income to charity but who never donates any of her own disposable income to charity. She genuinely believes that it is important to alleviate poverty in the developing world, but she also reasons that because she is working to increase awareness of the problem, she has no obligation to donate any of their own disposable income to the cause. Third, consider an animal rights activist who regularly eats cheeseburgers. When questioned about his behavior, he reasons that his individual acts of consuming animal products have no meaningful economic impact and do not affect how many animals are slaughtered. He advocates for preventing harm to animals at the collective level but does not change his behavior because he does not think his particular acts of meat consumption cause harm to any animals.

In all of these cases, the behavior and values of the people involved are logically consistent. But it does not follow from this fact that these are people of integrity. In fact, a more accurate appraisal is that they are exemplars of hypocrisy. The reason their behaviors are hypocritical is not because these people are acting illogically. Rather, the problem is that they are behaving in one way in the public or political sphere and in a radically different way with respect to their individual behavior. Integrity requires a stronger cohesion between behaviors in these different contexts.

The behavioral disconnect in these cases is also entirely self-serving, much like it would be in the context of reducing one's individual contributions to climate change. In each case, the person in question can advocate for a particular value or cause in the public sphere while simultaneously avoiding any serious personal sacrifice connected to that value or cause.[13] The priest is demanding that others respect the sanctity of monogamous marriage but is not willing to honor that value himself. The charity advocate wants others to donate their disposable income to an important cause but is not willing to make these donations herself. The animal rights activist campaigns for social change that would result in the end of meat-eating yet is not willing to voluntarily give up eating meat himself. This is not integrity. A person of integrity stands up for their values and does so even in the presence of temptations to abandon them. For a person of integrity, these values and causes are *always* important – not just important in a single context and certainly not just important when upholding those values and causes is convenient.

We should also wonder whether people who compromise their integrity in self-serving ways with respect to their individual behavior will stay committed to their political ideals. If they are willing to abandon their values in one sphere, what prevents them from doing the same in another sphere, especially if pursuing those political ideals becomes increasingly difficult or costly? Our principles and values, at least if they are deeply held, should not be contingent on whether they are personally convenient.

For the reasons sketched above, I hold claim (3) to also be true. The three premises then entail the argument's conclusion: individuals should reduce their personal greenhouse gas emissions. There are, of course, unresolved questions about how much one is required to cut their personal emissions to meet this obligation. The content of this obligation will be rather context-sensitive (Hedberg 2018a, pp. 71–73). It may vary depending on our professional obligations, the adequacy of public transportation in our communities, our personal dispositions, and a host of other factors that can make reducing emissions more or less difficult. An obligation to reduce emissions does not demand that we be environmental martyrs, but it does demand that we make genuine efforts to lower our personal carbon footprints with respect to those activities that are not connected to our personal survival.

So how does this integrity-based argument about personal carbon emissions relate to overpopulation and the morality of procreation? One obvious connection is that a duty to reduce personal carbon emissions entails an obligation to restrict one's procreative habits (Hedberg 2019). To reiterate a point from the previous section, having an additional child generates far more carbon emissions than any other activity a person can perform. Thus, if we are obligated to reduce our personal carbon emissions, there is no more plausible candidate for an activity to restrict than our personal procreation.

Additionally, an appeal to integrity can be used to argue directly for restrictions on procreation. We can outline that argument with only a few small alterations:

1★ Individuals should live with integrity.
2★ Individuals should cooperate in working toward a collective political solution to overpopulation.
3★ If individuals should live with integrity and cooperate in working toward a collective political solution to overpopulation, then they should limit their individual procreation.
4★ Therefore, individuals should limit their individual procreation. [1★–3★]

The first premise is unchanged from the prior argument, so it requires no new support. The second premise can be supported by the Population Reduction Argument from Chapters 3 and 4: it is clear that a collective international effort is required to respond adequately to population growth and its effects. The third premise has been altered, but the reasoning that underlies it is identical to the prior argument. Maintaining integrity requires the proper cohesion between the political commitment to facilitate a response to population growth and a personal commitment to limit one's own individual acts of procreation.

Both the duty not to contribute to massive, systemic harms and considerations tied to integrity point to a moral obligation to limit our family size. The question we now must answer is how many children we can permissibly have.

What are the limits of permissible procreation?

Children do not come in fractional increments, so the attempt to delineate personal procreative obligations must converge on a whole number – 0, 1, 2, 3, and so on. One initial reaction might be to think that the problem of overpopulation is so great that people should have no children, at least until the global population starts to decline. But one individual act of procreation will not have enough of an impact on population trends for an individual sacrifice of this magnitude to be demanded. For many, it would mean abandoning a central part of their life plans and foregoing a wide range of meaningful experiences.[14] Moreover, if we are thinking in terms of integrity, many people deeply value the creation of a biological family. This value is in tension with their commitment not to contribute to overpopulation, and so a compromise between these commitments should be struck. Maintaining integrity does not require abandoning one of these commitments wholesale in favor of the other. For some people, however, they may indeed be obligated not to procreate. Not everyone really wants to be a parent, and at present, there are compelling moral reasons to limit one's procreation. So in the absence of any compelling justificatory reasons to procreate, a person should have zero children.[15] In this case, foregoing procreation does not involve a serious sacrifice.

For many people, however, limiting their procreative activity does seem like a significant sacrifice. So how do we strike the proper balance between the demands on individual moral obligation and the role that procreation plays in people's life plans? For now, we have ruled out the possibility of a zero-child obligation. I think we can similarly rule out any upper limit above replacement fertility. Suppose a couple has three children – the equivalent of 1.5 children per person. This means that their actions will have the long-term consequence of *increasing* the global population. This action does not give proper moral weight to the obligation to limit one's procreation, especially since the justificatory reasons for having a *third* child are much weaker than the reasons that could be offered for having the first two. Having a first child involves a remarkable change in identity – a transformation from non-parent to parent – that does not occur with subsequent births. A second child creates the possibility of sibling relationships while still remaining below the replacement-level fertility rates. But what could justify going *above* replacement level fertility when these other valuable ends of procreation have already been achieved? Outside of unusual scenarios (e.g., becoming pregnant with triplets), I do not think there are any justifying reasons for performing this action.

If an upper limit of three children per couple is too permissive and a duty to forego all procreation is too strict, that leaves us with only two options for permissible procreation: one child per person or one child per couple. One assessment of these competing positions comes from Christine Overall (2012). She argues in favor of a one-child-per-person standard. She objects to a one-child-per-couple standard by appealing to the general demandingness of this obligation, the recognition that such a view does not permit all individuals to replace themselves, concerns about the effects of eliminating sibling relationships altogether, and worries about how a one-child norm could lead to sex selection (Overall 2012, pp. 181–183). Based on these considerations, Overall (2012) concludes that "an obligation to have only one child is at most supererogatory and unlikely to be sustainable" (p. 183). The more sensible position, she reasons, is to claim that adults have the responsibility to limit themselves to procreative replacement when their children will have large ecological footprints. Not all of Overall's reasons are persuasive, however.

Individual procreative obligations 95

Concerns about sex selection or the disappearance of sibling relationships are more salient concerns if we are considering a one-child policy that was heavily enforced, but I explicitly rejected such policies. In this chapter, we are assuming that the social policies in place are broadly permissive with respect to procreative choices and focusing on how individuals should assess what choices are morally permissible. Since a one-child-per-couple obligation would have a great deal of noncompliance in the absence of enforcement, these worries may not manifest in a meaningful way. Additionally, Sarah Conly (2016), a defender of the one-child-per-couple standard, would argue that sex selection should be countered by trying to address the root of the problem, which is a cultural bias in favor of boys (pp. 200–204), and that there is no consensus about whether only children are better or worse off than children with siblings (pp. 209–213). We need stronger reasons for thinking that the one-child-per-person norm would be problematic.

One of Overall's more unique arguments against a one-child-per-couple obligation is based on a right for individuals to replace themselves through procreation. Since her articulation of the argument is brief, I present the relevant passage here in its entirety:

> I suggest that a further problem with the one-child-per-couple obligation is that it implicitly negates one person in the couple. If a couple has two children, however, there is a child for each one – not in the sense that each raises only one child, but in the sense that each individual has replaced himself or herself. By contrast, a moral rule of only one child per couple says, in effect, "You ought not to replace yourself." (Perhaps it would also carry the message "You do not deserve to be replaced.").
> (Overall 2012, p. 182)

Overall's reasoning here is hard to decipher, but the core idea appears to be that there is something morally objectionable about the following moral principle: you ought not to replace yourself. The underlying thought may be that all people, merely by virtue of being rational and autonomous agents, have a right to replace themselves through procreation. Alternatively, the parenthetical remark may indicate the real problem – the concern that such a moral imperative would carry with it a pejorative message that should not be conveyed.

Like the prior arguments, this one is not compelling. The moral rule we are considering is contingent upon certain background conditions – namely, overpopulation and its ongoing effects – and so the moral rule of one-child-per-couple would say something more akin to "Under current conditions of overpopulation, not all people ought to replace themselves." This principle clearly doesn't connote the pejorative implication that people "do not deserve" to be replaced since it is derived from tragic social circumstances. The rationale for the principle has nothing to do with what people deserve. The notion that a person in a couple is "negated" when they are only allowed to have one child is also puzzling: it is not as if parents designate that a particular child is the father's replacement or the mother's replacement. Both parents claim the child as *theirs*. Why is it so important that parents be allowed to have enough children to ensure numerical replacement, particularly when the consequences of their doing so contribute to such significant detrimental consequences?

Overall later suggests that what is objectionable about a one-child-per-couple obligation is that it does not properly acknowledge the value of adult human beings. Although not directly stated, this claim can be inferred from what she says about the benefits of a one-child-per-person obligation:

> All persons get to (try to) have a child of "their own," if they want one, and the value of every adult is implicitly endorsed through the fact that each one is allowed to reproduce herself or himself. Such a responsibility implies that every person is sufficiently valuable to be worth replacing.
> (Overall 2012, p. 183)

Her suggestion then is that that a one-child-per-couple obligation would *not* imply that every person is sufficiently valuable to be worth replacing, but this obligation just does not entail this claim about the value of people's lives. In fact, the justification for this restriction on procreation arises directly from a concern about the value of people's lives – particularly, the negative value associated with widespread human suffering. In advocating a one-child-per-couple obligation, we would be claiming that the welfare loss of present and future people is so significant that we must limit our procreative activities. This reasoning does not imply making judgments about who is worth replacing and who is not.

Among the four reasons that Overall presents for favoring the one-child-per-person norm, only one remains – an appeal to the demandingness of limiting ourselves to one-child families. Is the duty to have only one child too demanding to be a moral requirement? I think the answer will depend on the particular people who we are considering. For many people, limiting themselves to a single child would not impose a significant hardship. First, to recall a statistic from Chapter 5, remember that 40 percent of pregnancies worldwide are unintended (Sedgh, Singh, and Hussain 2014). Those who would describe their pregnancy as mistimed might have wanted a child later, but a significant portion of births that resulted from these pregnancies were likely not an elaborate part of their parents' life plans. An even larger number of people procreate just to conform to social expectations or adhere to the lifestyle that they perceive as normal. Asking these people to limit themselves to a single child hardly seems like an unreasonable demand.

One common motive for having children is the belief that they will make their parents happy or at least be a general source of enjoyment. Despite the pervasiveness of this belief, social scientific research indicates that having children generally *decreases* happiness (e.g., Alesina, Di Tella, and MacCulloch 2004; Deaton and Stone 2014; Di Tella, MacCulloch, and Oswald 2003; Gilbert 2007, pp. 242–244; Hansen 2012; Margolis and Myrskylä 2015; Powdthavee 2008; Twenge, Campbell, and Foster 2003).[16] Raising children requires a massive investment of parents' time, money, and energy. As a result, having children generally puts a strain on people's marital satisfaction, life satisfaction, and self-reported happiness.[17]

It is not all bad for parents, however. There are benefits associated with having children. The trouble is that these benefits tend to accrue later in life – once the children have become adults – and it is unclear whether they ever aggregate sufficiently to overcome the sizeable drop in happiness that occurs immediately following the child's birth. One clearer benefit is that having a child generally increases the perceived meaningfulness of one's life despite diminishing one's well-being (Baumeister et al. 2013). Nevertheless, the central observation is that children do not generally make people happier than they would be without children. Those who think limiting their procreation to a single child is too demanding because doing so will deprive them of happiness are either unaware of the empirical research on the subject or in denial about its findings.

The strongest case for an exception the one-child-per-could standard would be prospective parents who genuinely want a child, have seriously reflected on their decision, who are excited about the experiences associated with gestation and parenthood, and who want their children to experience sibling relationships. Perhaps they have even considered adoption but encountered hurdles that render that difficult or impossible for them to pursue further. For these parents, would it be too demanding to ask them to refrain from having a second child? In this case and similar ones, it may be permissible for this couple to have a second biological child. They have the best motives possible for wanting a second child,[18] and limiting themselves to two total children keeps them below replacement-level fertility. I think it would be morally *better* for them to have only one child, but given that they are only replacing themselves through their procreation and given the importance of this second child to them, it may be unreasonable to demand that they remain from conceiving a second time. Even so, I suspect only a small portion of prospective parents will be able to meet the explanatory burden necessary to justify having a second child.[19] For many couples, the permissible number of children will be one.[20] And for people who have no serious interest in procreating, they morally ought to have zero children. We must now consider some potential objections to these conclusions.

Does the presence of reproductive rights provide moral justification for large families?

Sometimes, when people are told that they have a moral obligation to limit their procreation, their immediate thought is that this constitutes a violation of their rights. They might maintain that they have a right to have as many children as they want, and that any other standard infringes on their reproductive freedom in a morally objectionable way. I will return to this idea in Chapter 8, but for now, I want to point out that this appeal does not make sense in the context of individual moral obligations. Procreative rights are important in the context of developing social policies pertaining to procreation, but even if the right to procreate were unlimited, it does not follow from that fact that people are morally permitted to have an unlimited number of children (Overall 2012; Rieder 2016, pp. 51–52).

Most people acknowledge that we have a right to freedom of speech and expression. The ability to speak and express oneself freely is one of people's most critical and universal interests. Nonetheless, there are contexts where I can exercise this right in ways that are morally wrong, such as saying hurtful or slanderous things. I may be well within my rights to say those things, but speaking in ways that disrespect other people or that cause harm (e.g., by inciting violence) is still morally wrong. We might reason similarly about viewing sexually explicit material. My privacy rights may ensure that I be allowed to view such material, but it may still be wrong for me to do so. Perhaps the women in this material are objectified or exploited, or perhaps it reinforces gender stereotypes in ways that could adversely impact my moral character.

The key takeaway from these examples is that having the right to do something does not make doing it the right thing to do. Or, to put it another way, having *the right* to do X does not mean that one *ought* to do X. So if we apply this to the case of procreation, we can reason as follows: having the right to conceive as many children as you want does not mean that you ought to conceive as many children as you want. Rights are designed to protect people's fundamental interests, but they are not grounds for ignoring other moral obligations. If someone wants to reject the moral duty to limit procreation to one-child-per-person, an appeal to reproductive rights will not suffice.

Shouldn't environmentally conscientious people have more children?

One implication of my position is that even the most morally reflective and environmentally conscientious people should limit themselves to two children. Some may think that the environmentally conscientious are precisely the people who should have large families (Wisor 2009). Otherwise, the people who procreate more will generally be those who are not environmentally conscientious, and this could prove to be worse for the environment in the long term.

This objection has two fatal shortcomings. First, one of its underlying assumptions – namely, that there is a strong connection between the environmental views of parents and the environmental views of their children – is false. Many children reject the views of their parents or lack a clear understanding of them. As one illustration, in some studies, more than 50 percent of children either rejected to the political party affiliation of their parents or did not accurately perceive what that political affiliation was (Ojeda and Hatemi 2015). Human experience also offers bountiful other examples of children rejecting the values and beliefs of their parents. Short of indoctrination, we do not have compelling reasons to think that children will reliably adopt the beliefs and values of their parents.

Additionally, there are just better ways to raise the environmental conscientiousness of the general public. Rather than trying to raise children who have these values, we should educate those who already exist, develop social and cultural norms based on reducing overconsumption and living sustainably, and adjust our infrastructure to make it easier for people to live in environmentally friendly ways. Trying instead to create more environmentally conscientious people through procreation is foolish and counterproductive.

What about offsetting?

The most obvious strategy for reducing the ecological footprint tied to procreating is to limit one's procreation. An alternative strategy is to spend a considerable sum of money to reduce environmental impacts in order to compensate for procreating. The strategy of offsetting individual contributions to environmental impacts has been most commonly discussed in the context of climate change. Carbon offsetting, as it is called, is the process of either removing carbon from the atmosphere (e.g., by planting trees) or preventing the emission of carbon elsewhere in the world (e.g., by the development of wind turbines that generate renewable energy) as a means of offsetting the effects of one's own individual emissions. If done effectively, then for every unit of carbon that a person emits, they cause an equal sum of carbon to be either subtracted from the atmosphere or prevented from entering the atmosphere in the first place. The result is that the atmosphere contains the same amount of carbon as if that person had never emitted any in the first place.[21]

The idea behind offsetting is well-intentioned, and it is far better to try to offset one's environmental impacts than to do nothing. But from a practical standpoint, offsetting is not an adequate substitute for refraining from procreation. First, it will be incredibly difficult to get an accurate measurement of the overall ecological footprint tied to procreation so that one can determine how much to offset. Second, it will be difficult to appraise whether certain offsetting schemes actually make a difference. Suppose you donate money to help build a hydroelectric power plant, but the plant would have been built without your donation. In that case, your attempt to offset a portion of your ecological footprint failed because your action did not actually prevent any environmental impacts from happening. Or suppose you donate to an organization that aims to prevent deforestation in some region. If the land your money helped protect is one day deforested anyway, then in the long term, your attempt at offsetting failed. Third, many offsetting projects are appealing in large part because they are cheap, but if a significant number of people sought to offset the impact of their procreative activity by offsetting, that would quickly change, and it would be an option only for the rich. Thus, if we want a large-scale means of enabling people to fulfill their procreative moral obligations, offsetting will not work.[22]

Despite these drawbacks, offsetting could play a role in justifying the decision to have a second child. If parents were willing to attempt to offset some of the environmental impacts of having a second child, then that would make their explanation for having a second child more compelling – provided, of course, that they actually engage in the offsetting activities that they describe.

Policy measures and individual obligations

I have argued that couples who can exercise their procreative freedom should limit themselves to (at most) one child per person, especially if these people live in a nation with a large per capita ecological footprint. Those who can be satisfied with only a single child in their family ought to have only that one child, and people who have no significant interest in procreating should stay childless.

In the previous chapter, I proposed a variety of policy measures that could aid the long-term goal of population reduction. On the whole, these proposals aim to enhance procreative freedom rather than induce compliance through coercive measures. To make significant progress in decelerating population growth, we need to drop fertility at a rather swift pace. One may wonder how this fact accords with allowing some families to have two children. If people generally had two children, that would still lead to population growth in the short term as the young people currently living grow to reproductive age, and the population reduction that followed would be very gradual. Shouldn't I advocate for a moral standard that is more stringent so that progress is swifter?

This question gets at an important aspect of my position. At the policy level, the long-term goal is to reduce fertility well below two births per woman – perhaps to 1.5 births per woman or perhaps even lower. In this respect, individual obligations and policy goals diverge slightly. This is in part because children do not come in fractions. There is no way to neatly approximate the policy goal in individual behavior: couples who have no children or only one child will make the policy goal easier to achieve; couples who have two children will make the policy goal slightly harder to achieve. That does not mean that a couple who has two children has acted wrongly, however, since their particular case may be one where they are justified in having a second biological child.

These details notwithstanding, the purpose of specifying our moral obligations is not to appraise whether a specific couple is aiding or hindering the long-term goals with respect to population growth. After all, the policy goals are primarily meant to be achieved by creating conditions where people can exercise greater reproductive freedom and by creating a greater appreciation for and acceptance of small families. The point of assessing our personal procreative obligations is twofold. First, most of us value acting morally, and so we want to know what we should do to fulfill our moral duties. Second, the problems tied to overpopulation have created conditions where we need to subject our procreative decision-making to more serious scrutiny than we have in the past. Too many people still hold the belief that procreative decisions are personal and private in ways that exempt them from moral criticism. But this belief is false. Procreative choices have a significant impact on other people. Such choices are not private and warrant just as much moral scrutiny as other behaviors that can affect others' welfare. We must confront the moral challenges associated with procreating responsibly and think carefully about whether our procreative decisions are justified. The moral stakes are too high for us to neglect such questions.

Notes

1 There are many women who do not have the reproductive autonomy that is commonly enjoyed in some of the world's wealthiest nations. The arguments in this chapter do not apply to people who lack the freedom to choose their family size or who would suffer severe consequences (e.g., physical harm, ostracism) for trying to exercise such freedom.
2 For some expansion on this argument, see Nolt (2013b).
3 A further complication for this argument is the fact that a significant portion of emissions are structural in nature (Aufrecht 2011). That is, they are primarily attributable to the way that various infrastructures, such as our transportation systems and available housing, have been designed rather than deliberate choices made by individuals. Plausibly, the systems that direct the design of these infrastructures are more responsible for certain types of emissions than the individual actions of those influenced by these systems.
4 I borrow these examples from Rieder (2016, pp. 27–28).
5 Rieder also offers two other arguments for restricting procreation, which I have briefly appraised elsewhere (Hedberg 2018b). I focus on the duty not to contribute to massive, systematic harms because I consider it to be the strongest argument that Rieder presents.
6 Delineating every essential feature of integrity is challenging, but for some efforts in this regard, see Audi and Murphy (2006) and Scherkoske (2013).
7 Cognitive dissonance is a well-established psychological phenomenon in which the recognition of inconsistent beliefs or attitudes creates a feeling of discomfort. This discomfort usually motivates individuals to resolve the inconsistency. For the classic psychological studies on cognitive dissonance, see Festinger and Carlsmith (1959) and Festinger, Riecken, and Schachter (1964).
8 Hill (1979) presents this case as an act of symbolic protest and examines these acts from a deontological perspective. I present my integrity-based analysis as one viable explanation of why the woman's behavior is praiseworthy.

102 *Ethics, policy, and obligations*

9 Cheshire Calhoun (1995) even argues that this social aspect of integrity is its definitive feature.
10 This is a simplified version of the Integrity Argument that I defended in Hedberg (2018a, p. 67).
11 Not everyone agrees as to whether integrity can be possessed by evil or amoral people, but regardless of where one stands on that issue, having integrity is clearly *necessary* for being a morally good person. Without integrity, someone with all the correct moral beliefs could routinely act unethically by failing to live in accordance with their values. The possibility of evil people having integrity just suggests that possessing this trait is not *sufficient* for being a good person. This result should not be that surprising since many other virtuous traits (e.g., honesty, cleverness, loyalty) can be possessed by people whose overall character is quite vicious.
12 I thank Bob Fischer and Dan Shahar for bringing this objection to my attention.
13 There may be some extreme cases where a person is so invested in promoting a cause at the political level that they cannot change certain features of their personal life without compromising their political engagement. For instance, a climate activist who travels across the nation organizing rallies and giving public speeches may well have a higher individual carbon footprint than a normal citizen, but since these carbon emissions are connected with their political advocacy, I do not think these kinds of emissions would demonstrate a lack of integrity. Such a case is rare, though. Very few of us are invested in any political cause to such an extreme degree that we would be unable to integrate those political commitments into our personal lives.
14 See Dillard (2010) for a survey of the reasons that people value having children.
15 As elsewhere in this chapter, the obligation to limit procreation only applies to people who do not live in conditions where their reproductive autonomy is seriously diminished. There are many people who do not actually want to have children but are forced or pressured into doing so and could suffer serious harm if they acted otherwise.
16 Deaton and Stone (2014) think that the differences in happiness that they observed are negligible if one controls for parental choice. This would entail that parents who choose to have children are just as happy as people who choose not to have children.
17 I offer a more thorough analysis of the relationship between happiness and procreation in Hedberg (2019).
18 For some examination of the ways that motives affect the morality of prospective parents' procreative decisions, see Chambers (2019) and Lotz (2011).
19 Rieder (2016) echoes the same thought (p. 64).
20 It is worth acknowledging that parents who live in the developing world may be able to make a stronger case for having a second child than those in the developed world. Children born in the developing world will, at least in the short term, have smaller per capita ecological footprints than children born in the developed world, so the environmental impact of procreating in the developing world is not as large.
21 For some endorsements of carbon offsetting, see Broome (2012, ch. 5) and Spiekermann (2014). For some objections to carbon offsetting, see Bullock, Childs, and Picken (2009), Caney and Hepburn (2011), and Hyams and Fawcett (2013).
22 For further discussion of problems with carbon offsetting specifically, see Hedberg (2019).

References

Alesina, Alberto, Rafael Di Tella, and Robert MacCulloch. 2004. "Inequality and Happiness: Are Europeans and Americans Different?" *Journal of Public Economics* 88, no. 9–10: 2009–2042.

Almassi, Ben. 2012. "Climate Change and the Ethics of Individual Emissions: A Response to Sinnott-Armstrong." *Perspectives* 4, no. 1: 4–21.

Audi, Robert and Patrick Murphy. 2006. "The Many Faces of Integrity." *Business Ethics Quarterly*, 16, no. 1: 3–21.

Aufrecht, Monica. 2011. "Climate Change and Structural Emissions: Moral Obligations at the Individual Level." *International Journal of Applied Philosophy* 25, no. 2: 201–213.

Baumeister, Roy, Kathleen Vohs, Jennifer Aaker, and Emily Garbinsky. 2013. "Some Key Differences Between a Happy Life and a Meaningful One." *The Journal of Positive Psychology* 8, no. 6: 505–516.

Broome, John. 2012. *Climate Matters: Ethics in a Warming World*. New York: W.W. Norton & Company, Inc.

Bullock, Simon, Mike Childs, and Tom Picken. 2009. *A Dangerous Distraction: Why Offsetting Is Failing the Climate and People: The Evidence*. London: Friends of the Earth, 2009. www.foe.ie/assets/files/pdf/a_dangerous_distraction.pdf. Accessed December 12, 2019.

Calhoun, Cheshire. 1995. "Standing for Something." *Journal of Philosophy* 92, no. 5: 235–260.

Caney, Simon, and Cameron Hepburn. 2011. "Carbon Trading: Unethical, Unjust, and Ineffective?" *Royal Institute of Philosophy Supplements* 69: 201–234.

Chambers, Lindsey. 2019. "Wronging Future Children." *Ergo* 6, no. 5. DOI: 10.3998/ergo.12405314.0006.005.

Conly, Sarah. 2016. *One Child: Do We Have a Right to Have More?* Oxford: Oxford University Press.

Cripps, Elizabeth. 2011. "Climate Change, Collective Harm, and Legitimate Coercion." *Critical Review of International Social and Political Philosophy* 14, no. 2: 171–193.

Deaton, Angus, and Arthur Stone. 2014. "Evaluative and Hedonic Wellbeing among Those with and without Children at Home." *Proceedings of the National Academy of Sciences* 111, no. 4: 1328–1333.

Di Tella, Rafael, Robert MacCulloch, and Andrew Oswald. 2003. "The Macroeconomics of Happiness." *Review of Economics and Statistics* 85, no. 4: 809–827.

Dillard, Carter. 2010. "Valuing Having Children." *Journal of Law and Family Studies* 12: 151–198.

Festinger, Leon, and James M. Carlsmith. 1959. "Cognitive Consequences of Forced Compliance." *Journal of Abnormal and Social Psychology* 58, no. 2: 203–210.

Festinger, Leon, Henry W. Riecken, and Stanley Schachter. 1964. *When Prophecy Fails: A Social and Psychological Study of a Modern Group that Predicted the Destruction of the World*. New York: Harper and Row.

Fragnière, Augustin. 2018. "How Demanding is Our Climate Duty? An Application of the No-Harm Principle of Individual Emissions." *Environmental Values* 27, no. 6: 645–663.

Gilbert, Daniel. 2007. *Stumbling on Happiness*. New York: Vintage Books.

Hansen, Thomas. 2012. "Parenthood and Happiness: A Review of Folk Theories Versus Empirical Evidence." *Social Indicators Research* 108, no. 1: 29–64.

Hedberg, Trevor. 2018a. "Climate Change, Moral Integrity, and Obligations to Reduce Individual Greenhouse Gas Emissions." *Ethics, Policy & Environment* 21, no. 1: 64–80.

Hedberg, Trevor. 2018b. Review of *Toward a Small Family Ethic: How Overpopulation and Climate Change Are Affecting the Morality of Procreation*. *Kennedy Institute of Ethics Journal* 28, no. 4: 8–13.

Hedberg, Trevor. 2019. "The Duty to Reduce Greenhouse Gas Emissions and the Limits of Permissible Procreation." *Essays in Philosophy* 20, no. 1. DOI: 10.7710/1526-0569.1628.

Hill, Thomas. 1979. "Symbolic Protest and Calculated Silence." *Philosophy & Public Affairs* 9, no. 1: 83–102.

Hourdequin, Marion. 2010. "Climate, Collective Action, and Individual Ethical Obligations." *Environmental Values* 19, no. 4: 443–464.

Hyams, Keith, and Tina Fawcett. 2013. "The Ethics of Carbon Offsetting." *WIREs Climate Change* 4, no. 2: 91–98. DOI: 10.1002/wcc.207.

Jamieson, Dale. 1992. "Ethics, Public Policy, and Global Warming." *Science, Technology, and Human Values* 17, no. 2:139–153.

Jamieson, Dale. 2014a. "Climate Matters: Ethics in a Warming World by John Broome." *Ethics and International Affairs* 28, no. 2. www.ethicsandinternationalaffairs.org/2014/climate-matters-ethics-in-a-warming-world-by-john-broome/. Accessed November 8, 2019.

Jamieson, Dale. 2014b. *Reason in a Dark Time: Why the Struggle Against Climate Change Failed – and What It Means for Our Future*. Oxford: Oxford University Press.

Kawall, Jason. 2011. "Future Harms and Current Offspring." *Ethics, Policy & Environment* 14, no. 1: 23–26.

Lichtenberg, Judith. 2010. "Negative Duties, Positive Duties, and the 'New Harms.'" *Ethics* 120, no. 3: 557–578.

Lotz, Mianna. 2011. "Rethinking Procreation: Why It Matters Why We Have Children." *Journal of Applied Philosophy* 28, no. 2: 105–121.

Margolis, Rachel, and Mikko Myrskylä. 2015. "Parental Well-being Surrounding First Birth as a Determinant of Further Parity Progression." *Demography* 52, no. 4: 1147–1166.

Murtaugh, Paul, and Michael Schlax. 2009. "Reproduction and the Carbon Legacies of Individuals." *Global Environmental Change* 19, no. 1: 14–20.

Nolt, John. 2011. "How Harmful Are the Average American's Greenhouse Gas Emissions?" *Ethics, Policy & Environment* 14, no. 1: 3–10.

Nolt, John. 2013a. "Replies to Critics of 'How Harmful Are the Average American's Greenhouse Gas Emissions?'" *Ethics, Policy & Environment* 16, no. 1: 111–119.

Nolt, John. 2013b. "The Individual's Obligation to Relinquish Unnecessary Greenhouse Gas-Emitting Devices." *Philosophy & Public Issues* 3, no. 1: 139–165.

Ojeda, Christopher, and Peter Hatemi. 2015. "Accounting for the Child in the Transmission of Political Party Identification." *American Sociological Review* 80, no. 6: 1150–1174.

Overall, Christine. 2012. *Why Have Children?* Cambridge, MA: MIT Press.

Powdthavee, Nattavudh. 2008. "Putting a Price Tag on Friends, Relatives, and Neighbours: Using Surveys of Life Satisfaction to Value Social Relationships." *Journal of Socio-Economics* 37, no. 4: 1459–1480.

Rieder, Travis. 2016. *Toward a Small Family Ethic: How Overpopulation and Climate Change Are Affecting the Morality of Procreation.* Cham, Switzerland: Springer.

Scherkoske, Greg. 2013. "Whither Integrity I: Recent Faces of Integrity." *Philosophy Compass* 8, no. 1: 28–39.

Schwenkenbecher, Anne. 2014. "Is There an Obligation to Reduce One's Individual Carbon Footprint?" *Critical Review of International Social and Political Philosophy* 17, no. 2: 168–88.

Sedgh, Gilda, Susheela Singh, and Rubina Hussain. 2014. "Intended and Unintended Pregnancies Worldwide in 2012 and Recent Trends." *Studies in Family Planning* 45, no. 3: 301–314.

Spiekermann, Kai. 2014. "Buying Low, Flying High: Carbon Offsets and Partial Compliance." *Political Studies* 62, no. 4: 913–929.

Twenge, Jean, W. Keith Campbell, and Craig Foster. 2003. "Parenthood and Marital Satisfaction: A Meta-Analytic Review." *Journal of Marriage and Family* 65, no. 3: 574–583.

Wisor, Scott. 2009. "Is There a Moral Obligation to Limit Family Size?" *Philosophy & Public Policy Quarterly* 29, no. 3/4: 26–31.

World Health Organization. 2009. *Global Health Risks: Mortality and burden of disease attributable to selected major risks.* Geneva, Switzerland: WHO Press. www.who.int/healthinfo/global_burden_disease/GlobalHealthRisks_report_full.pdf. Accessed December 6, 2019.

Wynes, Seth, and Kimberly Nicholas. 2017. "The Climate Mitigation Gap: Education and Government Recommendations Miss the Most Effective Individual Actions." *Environmental Research Letters* 12, no. 7. http://iopscience.iop.org/article/10.1088/1748-9326/aa7541. Accessed February 18, 2020.

Part III

Objections from alternative approaches to procreative ethics

7 Antinatalism

I have argued that we have a moral obligation to reduce population growth in response to escalating environmental problems. I have also assessed what policy measures could be adopted to meet this obligation and what considerations should guide individuals in their procreative decision-making. I will now consider two objections to my position. According to one of these objections, my position is too permissive with respect to procreation and allows people to have too many children; according to the other, my position is too restrictive with respect to procreation and does not adequately respect reproductive rights. For now, I will focus on the concern that my moral standards for procreation are too lax.

Antinatalism is the philosophical position that it is morally bad to be born. Rather than viewing life as something to be cherished or savored, antinatalists tend to view life as harmful and view the imposition of life on another to be morally unacceptable. This outlook on existence strikes most as baffling because it contradicts so many values and social norms that are typically associated with procreation. Nevertheless, despite its counterintuitive nature, antinatalism has been gaining traction in some academic circles. As an illustration, consider this except from the opening paragraph of Jamie Nelson's (2016) review of *Permissible Progeny? The Morality of Procreation and Parenting*:

> [T]he tenor of the text is that, given environmental fragility and existing children in need of families, bearing and begetting as such are morally fraught enterprises, certainly on the defensive, and possibly best seen as indulgence in an expensive hobby. Antinatalism is taken in general quite seriously throughout, sometimes bracketed, but never directly confronted.

Antinatalism also surfaces occasionally in non-academic contexts, such as when Raphael Samuel declared his intention to sue his parents for giving birth to him without his consent (Pandey 2019).[1]

Antinatalism is not a popular position among the general public, so it might seem unusual to address it at length. But there are actually quite a few arguments that would support an antinatalist outlook on procreation, and a view's lack of popularity does not mean that it is wrong. We have to look at the arguments for and against it to establish that conclusion. Moreover, if antinatalists are correct, then we might be morally required to pursue policy measures that restrict procreation as much as possible, which would result in a much different outlook on population policy than what I proposed in Chapter 5. Our long-term goal would not be reducing population to a sustainable level but bringing the human population down to zero.

I distinguish between two types of antinatalism. One view, which I will call Unconditional Antinatalism, holds that procreation is morally bad in *all circumstances*. A different view, which I will call Conditional Antinatalism, holds that procreation is morally bad under *present circumstances*. Conditional Antinatalism acknowledges that procreation could be permissible – or even morally good – if our circumstances were to change in the future. I start by examining the major arguments in favor of Unconditional Antinatalism.

Benatar's unconditional antinatalism

The philosopher who has been the most prolific and outspoken defender of Unconditional Antinatalism is David Benatar (2006, 2015). Benatar has put forward three distinct arguments for antinatalism in his writings. I will appraise each of them separately.

The axiological asymmetry

His first argument is based on an axiological asymmetry between existence and non-existence. This argument relies on four claims:

1. The presence of harm is bad.
2. The presence of benefit is good.
3. The absence of harm is good, even if that good is not enjoyed by anyone.
4. The absence of benefit is not bad unless there is somebody for whom this absence is a deprivation.

(Benatar 2015, p. 23)

If you accept all four of these claims, then nonexistence appears to always have an advantage over existence. That probably is not obvious from looking at these claims in the abstract, so I will unpack the reasoning a little further.

If you exist, then it is inevitable that you will be harmed at some point; if you do not exist, then you cannot be harmed. These points should be uncontroversial, so we should acknowledge that this is a consideration in favor of nonexistence. But if you exist, then you can be benefited (e.g., by having good experiences), and if you don't exist, you cannot be benefited. That seems like a significant drawback, but Benatar actually disagrees. He does not think there is anything bad about not being benefited when no one exists to experience the benefit, but he thinks it is good to avoid harm even when no one exists to experience that good. So, according to his reasoning, nonexistence has the positive feature of avoiding harm and no negative feature associated with missing out on its benefits. In contrast, existence has the negative feature of being harmed and the positive feature of being benefited. On this assessment, existence has positive and negative aspects while nonexistence has one positive aspect (i.e., avoiding harm) and no negative aspects. So Benatar holds that nonexistence is morally preferable to existence: on his view, the harms imposed by existence cannot be justified.

Benatar's asymmetry argument has garnered little sympathy but plenty of critical attention (e.g., Bayne 2010; Belshaw 2007; Bradley 2010; Brown 2011; DeGrazia 2010; Harman 2009; Kaposy 2009; Overall 2012; Smuts 2014).[2] There are many available objections, but I will mention the two that I consider the most decisive. First, I see no compelling reason to accept claim (3) in Benatar's list: when no one exists to be harmed or benefited, both the absence of harm and the absence of benefit seem neutral. Why regard either of these states of affairs as good or bad when they are not good or bad for any existing entity? Second, even if we were to accept all of Benatar's claims, it is still possible to resist his conclusion. As Belshaw (2007) highlights, Benatar assumes that a package containing something good and something neutral is preferable to a package containing something good and something bad. Whether this is true depends on the quantity of goods and bads that are part of the equation. If I am choosing between a free sandwich and a four-course meal that costs one dollar, I might well opt for the four-course meal. Picking the free sandwich gives me a small benefit – namely, the sandwich – and monetary cost is neutral: since the sandwich is free, I am neither negatively nor positively affected with respect to my finances. The four-course-meal provides me with a much larger benefit but at a small cost. Even so, choosing the four-course-meal in this scenario hardly seems irrational because the benefits are great and the costs are small. We might well assess existence the same way: there are some costs associated with being alive, but generally these costs are outweighed by the benefits sufficiently as to make existence preferable to nonexistence. That line of reasoning takes us to Benatar's second argument for antinatalism.

The quality-of-life argument

Benatar directly challenges the assumption that human lives are generally good and worth living. In a nutshell, Benatar (2006, pp. 60–92, 2015, pp. 40–77) argues that quality of life for the majority of human beings is very bad. Because human lives are usually very bad, we are not justified in creating new people and subjecting them to this fate, provided that we can avoid doing so. Any act of procreation is too risky because the odds are overwhelming that the person born will live a very bad life.

Benatar's primary support for this claim is based on empirical evidence that we overestimate the quality of our lives. He surveys various psychological phenomena that cause us to see our lives as being better than they objectively are (Benatar 2006, pp. 64–69, 2015, pp. 41–54). One of these is optimism bias – the general tendency of human beings to interpret our experiences in an optimistic fashion. This bias manifests when we, for example, tend to remember a greater number of positive events in our lives than negative events or when we overestimate how good events in the future will be.[3] Another is adaptation (or habituation), which is our tendency to adjust our expectations to suit our circumstances. A pronounced example of this occurs in cases where those who become paraplegic often become happy again, according to their own self-reports, within one year after losing the use of their legs (Brickman, Coates, and Janoff-Bulman 1978). A third phenomena that leads to us overestimating the quality of our lives is our tendency to make implicit comparisons between ourselves and others (Wood 1996). Since we often compare ourselves to those around us to assess how well our lives are going, widely shared negative features of human life are typically overlooked in our assessments of our own well-being. Moreover, we tend to compare ourselves with people who are worse off than we are (Brown and Dutton 1995), further biasing our assessments in an optimistic direction.

Optimism bias, adaptation, and social comparison compound to make us significantly overestimate the quality of our lives. On this point, Benatar and I have no disagreement: the empirical evidence for these tendencies is substantial, and they make sense from an evolutionary perspective. Those with a broadly optimistic outlook on their lives will, other things equal, be more likely to survive and reproduce than those who are more pessimistic.[4] The problem for Benatar is that we can accept these claims without accepting his conclusion: acknowledging that we generally overestimate how good our lives are does not entail that our lives are so bad that no one should be born. Suppose that a life with a rating of 100 is a blissful life, and a life with a rating of 0 (or lower) is a truly awful life. If I overestimate my life as being a 70, then my life might still be pretty good if it is evaluated from a purely objective lens. Maybe my life is closer to a 40 or a 30. That's not exceptional but still decent – a life well worth living.

For Benatar's argument to work, we need an explanation for why the gap between the perceived quality of our lives and their actual quality is as enormous as he suggests. To defend his view, Benatar appeals to the criteria by which we might judge lives as good or bad. According to an influential taxonomy of views concerning the quality of life, there are three accounts of what makes a life go well or poorly.[5] On hedonistic theories of well-being, lives fare well or poorly depending on the quantity of pleasure and pain that is experienced. On desire-fulfillment theories of well-being, lives fare well or poorly depending on the extent to which a person's desires are fulfilled. Finally, on objective list theories of well-being, lives fare well or poorly to the extent that they contain certain things that are objectively good or bad.

Some items on the objective list are good or bad for one's life independent of their connection to pleasure and pain or to the person's desires. Benatar attempts to show that our lives fare poorly on all three of these theories.

His general strategy is to highlight the various negative aspects of our lives that we routinely minimize or overlook. From the hedonistic perspective, these take the form of minor pains and discomforts, such as hunger, thirst, allergies, headaches, nausea, and boredom (Benatar 2006, pp. 70–72). We are fairly familiar with the great tragedies that can befall human lives – early death, cancer, depression, and other chronic or life-threatening ills – but these more banal pains sometimes go unnoticed. He also claims that pleasures tend to be short-lived while pain and discomfort are often long-lasting (Benatar 2015, pp. 48–49). With respect to desire-fulfillment, Benatar (2006) notes that a much larger share of our lives is characterized by unsatisfied desires than satisfied ones (p. 74). As soon as we fill one desire, another usually rises to take its place, so desire frustration tends to be a more common experience than desire fulfillment. Finally, from the perspective of objective list theories, Benatar denies that anyone lives well from the point of view of the universe: our lives are fleeting and characterized by far too much pain and discomfort to be considered good when compared to a more idealistic possible life, such as that of a being that lives almost pain free for 300 years.

I consider Benatar's assessment of how people fare according to objective list theories to be thoroughly implausible. He sets the standard for a good human life so outlandishly high that no one can reach it, but all he succeeds in showing is that some possible beings could have *better* lives than human beings do. That does not establish that human lives are *bad*. It might be true that dogs might live better lives if they could enjoy the pleasures associated with reading a good novel, but no one would ever judge a dog's life as bad because they did not experience those pleasures. Similarly, a human life might be better if it extended to 300 years, but it is ridiculous to claim a human life is bad because it does not last that long.

Benatar's assessment of the quality of our lives does not fare much better from the perspective of hedonistic and desire-fulfillment theories. One significant factor that Benatar overlooks is that many of the minor pains he mentions are either balanced by other feelings and sensations (such that we are not in constant states of discomfort) or even pleasant in certain contexts (Wasserman 2015, pp. 156–157). While it would be very bad to be in a constant state of hunger, it is hardly bad to feel hungry right before a large meal. The anticipation of satisfying that hunger and the actual satisfaction that follows can be far more pleasurable than that of a meal eaten on a partially full stomach. In similar fashion, feelings of minor discomfort are often nullified or entirely overridden by minor pleasurable sensations that we routinely fail to notice. Suppose, for instance, that the weather outside is a bit hotter than we would prefer, but that the surrounding greenery is also aesthetically pleasing to us. In practice, because both phenomena are commonplace, we might not notice either of these features of our surroundings.

In the prior paragraph, I illustrated a representative example of the central shortcoming in Benatar's efforts to appraise the quality of our lives. Since he is making an empirical argument, he needs to explain – in detail – all of life's good features and how they compare to all of life's bad features (Marsh 2014, p. 447). Benatar has not undertaken this task: there is no serious effort to catalog all the minor pleasures we routinely experience or grander pleasures and satisfactions (e.g., from accomplishing goals or completing major life projects). If we are to conclude that our lives go badly on the whole, then an in-depth comparison of the good and bad features of our lives must be provided.[6]

An additional complication in Benatar's calculation is that some things people value cannot be neatly explained in terms of pleasurable and painful mental states. People usually want their lives to be meaningful, for example, but having a meaningful life is not equivalent to being happy or to having a certain portion of one's mental states be positive in nature. In fact, there is evidence that higher levels of worry, anxiety, and stress correlate with higher levels of meaningfulness (Baumeister et al. 2013). Those who center their lives around substantial and difficult projects – which will often be a deep source of meaning for the person pursuing them – are more likely to experience these unpleasant mental states, but that does not mean that they would rather pursue a different life plan or that they are acting irrationally. Rather, this phenomenon highlights how there is more that we care about in our lives than just the aggregation of our positive and negative mental states. Insofar as a hedonistic theory cannot properly take these other valuable components of our lives into account, it cannot accurately assess how good a person's life is.

The lack of empirical rigor in Benatar's assessment also undermines his appeal to desire-satisfaction theories. Ordinarily, the way we would appraise how well a life goes on a desire-satisfaction account would be to see how many desires the person has and how many of them are ultimately satisfied. Benatar does not undertake this task in an empirically informed way: there is no effort to gather empirical evidence on how many desires people typically have or how many of them are typically fulfilled in a lifetime. He does highlight some of the common unfulfilled desires that people experience, but plenty of desires are also routinely fulfilled each day. How are we to judge whether fulfilled desires outnumber unfulfilled desires? More to the point, could such a comparison even be done? There may be no way to individuate desires in a way that is informative and not arbitrary. The desire to finish writing a novel, for instance, could be understood as a single desire (e.g., to finish the book in its entirety) or as a series of smaller desires (to finish chapter 1, to finish chapter 2, etc.). Which of these portrayals is accurate? How many desires are in play here? If there is no non-arbitrary way to answer these questions, then there is no fact of the matter regarding the *number* of desires that we have or the proportion of our desires that are satisfied or unsatisfied. Overall, Benatar's attempts to conclude that our lives are generally bad do not succeed, and so they do not provide strong grounds for endorsing antinatalism.

The misanthropic argument

While Benatar's earliest work on antinatalism focused on defending his axiological asymmetry and his judgments about the quality of human life, he has recently advanced a different argument in favor of antinatalism. Rather than trying to establish that procreation harms the person who is created, this argument tries to establish that procreation is wrong because of the harm it causes to other people. Benatar (2015) calls this argument "misanthropic" because it focuses on "the terrible evil that humans wreak, and on various negative aspects of our species" (p. 78). Here is how the Misanthropic Argument proceeds:

1. We have a (presumptive) duty to desist from bringing into existence new members of species that cause (and will likely continue to cause) vast amounts of pain, suffering, and death.
2. Humans cause vast amounts of pain, suffering, and death.
3. Therefore, we have a (presumptive) duty to desist from bringing new humans into existence.

(Benatar 2015, p. 79)[7]

The argument has valid form, and the second premise is obviously true. Even the most optimistic person must acknowledge that human beings often commit moral atrocities, and our history of violence, oppression, exploitation, and deception provides plenty of evidence for the claim. Thus, the soundness of the argument hinges entirely on the first premise. Is this claim true?

Benatar thinks that this premise would be widely accepted if the species under consideration were not human. He asks us to imagine people breeding a destructive species of nonhuman animal or scientists releasing a deadly virus and argues that both of these practices would be widely condemned (Benatar 2015, pp. 101–102). Since we hesitate to reach the same verdict about human beings, he suggests our resistance is only due to a bias in favor of our own species. But Benatar makes two mistakes here. First, he is wrong about our judgments about other species. Many predator species cause a great amount of suffering to other animals: they have to savagely kill other animals for their own sustenance. Yet there is not a widespread condemnation of these species or a strong public outcry for the elimination of predation. In fact, we sometimes undertake efforts to re-introduce predator species into environments where their numbers have dwindled. Predators do not cause as much harm as human beings – in large part because their numbers are so much smaller – but these observations provide some evidence that not everyone would immediately accept Benatar's judgment about the first premise even when it concerns nonhuman species.

116 *Approaches to procreative ethics*

The second mistake is that Benatar overlooks the fact that human beings also perform actions that are morally good. No one is morally perfect, but few are as monstrous as the murderers and animal abusers that he cites as examples of human evil. Surely the good that people do counts in favor of creating more of them, so we need further discussion of just how great the harms of the typical person are and how they compare to the good that the person does. Furthermore, there is some evidence that human beings are improving. We are, for instance, much less prone to violence than we were in the past (Pinker 2011), and there have been growing cultural trends in the developed world of acceptance of people of different races, nationalities, genders, sexual orientations, and religions.[8] These trends suggest that humanity is morally improving, even if the improvement is slower than we might prefer.

A further problem for Benatar's position is that his observations about collective human behavior have little import for individual procreative decision-making. One reasonable response to Benatar's long list of human evils is to instill parents with procreative caution: they should be reflective and determine whether they can provide their child with the proper upbringing and education so that their children will be extremely unlikely to commit, or be complicit in, dreadful wrongdoing (Wasserman 2015, pp. 167–168). Many parents can meet this obligation, and for them, it is not wrong to procreate. Benatar could well be right that there are some fairly significant restrictions on permissible procreation and that many people violate them, but that only tells us what we probably already suspected. Parents have very strong obligations to their children. Not everyone is able to raise children effectively, and even for those who are able to be good parents, their life circumstances may only be conducive to raising children well at certain times or under certain financial conditions. Still, many people can meet these conditions, and so Benatar's prohibition on procreation is not as broad as he argues.

Other arguments for unconditional antinatalism

Benatar is not the only person to offer recent arguments against the morality of procreation. In this section, I examine arguments that procreation is wrong from several other authors.

Shiffrin's consent-based argument

Seana Shiffrin (1999) argues against the permissibility of procreation on the basis of two observations. First, coming into existence renders a person vulnerable to a wide range of harms. Second, a person cannot consent to coming into existence. Often, it is morally wrong to cause harm to someone else or to expose them to risk of harm unless they give consent to the action in question (e.g., before undergoing surgery). If consent is indeed required to nullify the harms that accompany continued existence, then procreation seems morally impermissible.

This reasoning proceeds too quickly, however. After all, we are sometimes justified in causing harm to someone when it prevents them from suffering a greater harm, particularly when it is impossible or unrealistic to acquire consent. If a speeding driver fails to stop at a red light when a pedestrian is crossing the street, I am justified in shoving the pedestrian to the ground if it is the only way to get him out of the driver's path even though obtaining his consent to this action is not possible and he may suffer minor physical injuries because of my action. But Shiffrin (1999) claims that this justification does not hold when we harm someone merely to provide the person harmed with a benefit:

> Absent evidence that the person's will is to the contrary, it is permissible, perhaps obligatory to inflict the lesser harm of a broken arm in order to save a person from significant greater harm, such as drowning or brain damage from oxygen deprivation. But, it seems wrong to perform a procedure on an unconscious patient that will cause her harm but also redound to her a greater, *pure* benefit. At the very least, it is much harder to justify. For example, it seems wrong to break an unconscious patient's arm even if necessary to endow her with valuable, physical benefits, such as a supernormal memory, a useful store of encyclopedic knowledge, twenty IQ points worth of extra intellectual ability, or the ability to consume immoderate amounts of alcohol or fat without side effects. At the least, it would be much harder to justify than inflicting similar harm to avert a greater harm, such as death or significant disability.
>
> (p. 127)

Beyond these examples, Shiffrin also suggests that it would be wrong for a wealthy islander to air drop gold bars into a neighboring island community when doing so breaks someone's arm, even when this increased wealth makes the islanders (including the victim) better off overall and even when there are no other viable means of transporting his gold to this community.

Shiffrin may be right to stress the moral seriousness of imposing harm on others or even just exposing them to likely harms, but her supporting examples are problematic if we try to extend them to the case of procreation. The harm in these cases – breaking a person's arm – constitutes a rights violation.[9] Procreation, in contrast, does not involve any clear rights violations. For this reason, breaking a person's arm requires a particularly strong moral justification, but it remains unclear whether procreation requires a comparably strong moral justification, particularly when we acknowledge that we routinely expose our children to potential harms to provide them with pure benefits (DeGrazia 2012, p. 153). When we make our children play outside on a sunny day, we expose them to potential harms – bruises, cuts, splinters, and so on – that they would be much less likely to suffer if they stayed in a carpeted indoor environment, but we assume that whatever benefits come from being outdoors outweigh the risks of these harms. When we send our children to school, we know there is a chance they will be mocked, ridiculed, humiliated, or otherwise hurt by their peers, but we assume that the benefits associated with making friends and getting an education are worth the risk of suffering these harms. For these reasons, we should reject Shiffrin's claim that it is always wrong to expose someone to the risk of harm to provide pure benefits when the person affected cannot consent.

Jimmy Licon's consent-based argument

Jimmy Licon (2012) offers an alternative consent-based argument against the morality of procreation. His argument can be outlined as follows:

1   An individual is justified in subjecting someone to potential harm only if either: (a) they provide informed consent, (b) such is in their best interests, or (c) they deserve to be subjected to potential harm.
2   Bringing someone into existence is potentially subjecting them to harm.
3   Individuals that do not exist: (a) cannot give their consent to being brought into existence, (b) do not have interests to protect, and (c) do not deserve anything.
4   Hence, procreation is not morally justified.

(Licon 2012, p. 88)

While this argument does avoid appealing to Shiffrin's principle about the wrongness of causing harm to bestow pure benefits, the argument on the whole is not any better than hers.

The central flaw in the argument is in its third premise. Licon claims that individuals who do not exist lack interests. As stated, this premise is false. The phrase "individuals who do not exist" could be interpreted to refer to future people, possible people, or both future people and possible people. The premise is only true if it is restricted to referring to possible people. Recall from previous chapters that I use the term "future person" to refer to someone who *will* exist later even though they do not at present. A merely possible person is, in contrast, someone who *could* exist but never actually will exist. Merely possible people do not have interests, but future people *do* have interests. At a minimum, they have interests in the basic requirements for a decent human life. Indeed, much of our way of talking about future generations operates on the assumption that they have interests that we can prevent or thwart. There is nothing incoherent in saying that someone's grandchildren, whomever they are, have an interest in a getting a good education, and there is nothing incoherent in saying that people living in 2100 have interests in clean air and a reliable supply of fresh water. Children who are not yet born but *will* be born are best understood as future people, and so they have interests that we can promote or thwart. When we promote these interests sufficiently well, doing so can justify exposing them to potential harms.

Perhaps Licon could respond by pointing out that while future people *will* have interests, they do not have interests *yet*. When we speak of our children having interests when they are not yet born, we may be misrepresenting what we really mean. Even if we can grant Licon this move, it does not salvage his argument. On this interpretation, the argument is invalid: we can affirm all three premises but deny the conclusion because future people *will* have interests to protect. Sufficiently protecting or promoting those interests is enough to justify the risk of harm, particularly when the probability of significant harm is extremely low and the probability of beneficial experiences is extremely high. This thought accords not only with my remarks in the prior paragraph but also with widespread considered judgments about when procreation is justified. In fact, the claim that procreation is justified when the child would have an excellent chance of living a good life is one of the most stable and widely shared judgments in the ethics of procreation. To make his argument valid, Licon would have to alter part (b) of the third premise to read "do not have interests to protect and will not have interests to protect." But this construction renders the premise clearly false, since future people will certainly have interests to protect. Thus, regardless of which change is made, the argument fails.

Häyry's Risk Aversion Argument

Matti Häyry offers a further argument against the permissibility of procreation. When it comes to procreative decisions, he endorses the maximin rule, a principle of reasoning endorsed by John Rawls (1999, §26–28) in the context of his political philosophy. The maximin rule states that in some situations where probabilities of specific outcomes are uncertain, we ought to choose the alternative in which the worst outcome is superior to the worst outcomes of the alternatives. In other words, we seek to minimize our potential losses.

Häyry argues that the maximin rule should be applied to reproductive decision-making because all lives carry a risk of being worse than having not been born at all, at least from the perspective of the person living such a life. The risk may be small, but it is always a possibility. Since we always have the option of refraining from procreation, we can always avoid this disastrous outcome, and Häyry (2004) thinks that is precisely what we should do:

> When people consider the possibility of having children, they confront the following choice. They can decide not to have children, in which case nobody will be harmed or benefited. The value of this choice, in terms of potential future individuals and their lives, is zero. Alternatively, they can decide to have children, in which case a new individual can be born. If this happens, the life of the future individual can be good or bad.

120 Approaches to procreative ethics

> The eventual value of the decision, depending on the luck of the reproducers, can be positive, zero, or negative. Since it is rational to avoid the possible negative outcome, when the alternative is zero, it is rational to choose not to have children.
>
> (p. 377)

If we genuinely believe that the rational and morally appropriate course of action is to avoid the worst possible outcome in this scenario, then Häyry is right: we should refrain from procreation. But is the maximin rule really the appropriate decision procedure under these circumstances?

Rawls (1999) notes that the maximin rule "is not, in general, a suitable guide for choices under uncertainty" (p. 133). He then specifies three conditions that must be met for an appeal to the maximin rule to be appropriate:

1   The probabilities of the possible outcomes are unknowable, or there exists some reason for discounting the estimated probabilities.
2   The person choosing cares very little about what she might gain above the minimum; it is not worth taking a chance to try to gain a further advantage.
3   The worst possible outcome is one that the person making the choice cannot accept – typically one that involves a grave risk.

In the case of procreation, only the third condition is met. Often we have strong reasons to believe that a child will have a very high probability of living a good life; in other cases, we have strong reasons to believe that a child's life prospects are much worse (e.g., because of a debilitating genetic condition). Thus, we can form reasonable estimates about the probabilities of the possible outcomes. Furthermore, although not-yet-existing children cannot choose to be born, it is clear that they have something substantial to gain from taking the gamble – all their positive future experiences. So, if they could make a choice, they might well opt for the gamble, even knowing that there was a small probability of a disastrous outcome.[10]

As these observations suggest, Häyry's reasoning is excessively risk-averse. We do not apply maximin-style reasoning very often, even in scenarios involving life-and-death risks. Many actions, including those as common as driving cars, impose unlikely but severe risks on others. Moreover, in the realm of policy, we routinely avoid choosing the safest policy. A 20 mile-per-hour speed limit on highways would surely prevent many fatalities, but would anyone endorse such a policy? Even though the harms caused by higher speed limits are severe, many believe that the benefits offered by the higher speed limits outweigh the costs associated with the harms. The mere possibility of severe harm does not provide a strong enough reason to categorically avoid an action that may result in that outcome, and so we do not have an obligation to avoid all procreation. Our duty is to minimize the risk of serious harm to our children *once they are born*, not to forego procreation altogether.

Arguments for conditional antinatalism

The preceding arguments for antinatalism aimed to establish that procreation is always wrong – that people should avoid procreating regardless of their circumstances. Some other arguments for antinatalist conclusions are not as extreme and aim only to establish that antinatalism is true given current empirical circumstances. This section engages with two of these arguments.

The duty to adopt

Due to the number of children who already exist and lack parents, Daniel Friedrich (2013) argues that some people are under a moral obligation to adopt children rather than procreating. His argument rests on an empirical observation and a moral principle. The empirical observation is that, for those of us who want to be parents, "we can protect parentless children from serious harm at little cost to ourselves by adopting them" (Friedrich 2013, p. 25). The moral principle Friedrich proposes is the claim that we ought to protect other people from serious harm when we can do so at little cost to ourselves. These two claims, if true, generate the conclusion that some people, if they are going to undertake the task of raising children, have a moral obligation to adopt children rather than creating their own.[11]

Friedrich bolsters his case by considering a wide array of objections to his position. Most of these objections stem from false beliefs about adoptable children, such as claims that adoptable children are more likely to be maladjusted or have behavioral problems or that parents generally cannot love adopted children as much as their own biological children (Friedrich 2013, pp. 28–31). Friedrich rightly points out that many would probably be more willing to adopt if they were to abandon these false beliefs and give appropriate weight to the upsides associated with adoption. But he also makes a concession that threatens to undermine the argument's significance. For certain people, the experiences associated with pregnancy and childbirth as well as other aspects of having biological children (e.g., family resemblance) are significant parts of their life plans, and this fact will not change even after full consideration of all the information concerning the choice to adopt rather than procreate.[12] Friedrich (2013) states that such people are exempt from the duty to adopt (p. 31). While there is no way to know how many people satisfy this criterion, this admission demonstrates that the duty to adopt may only be applicable to a relatively narrow range of people. After all, biological and cultural factors incline people to prefer procreation over adoption, and people often structure their lives using the creation of a biological family as a focal point. Thus, a great many people may be beyond the scope of Friedrich's proposed duty to adopt.

A further limitation of Friedrich's argument is that there are only so many children in the world in need of adoption. Some have argued that the number of people seeking to adopt children is significantly greater than the number of children who have been identified as requiring adoption – in both the western and non-western world (e.g., Cantwell 2003; Graff 2009; Lammerant and Hofstetter 2007, pp. 4–5). In other words, there already appear to be more parents interested in adoption than there are children available to adopt. Friedrich (2013) notes that the number of children identified as *adoptable* is likely much lower than the number of children who are actually in need of adoption (p. 34). Many countries lack the resources or cultural environment needed to maintain institutions that could properly identify children who need to be adopted, and as a result, the number of children who need to be adopted is probably much higher than the number presently available for adoption. Friedrich is surely right about this: UNICEF (2018) estimates that there were 140 million orphans in 2015. The problem, of course, is that until those children actually *are* available for adoption, it is not possible for would-be parents to adopt them and therefore implausible to suggest that these prospective parents have a duty to do so.[13] Some parents will still be in a position where local children are in need, but the number of available children to adopt significantly limits the scope of any duty to adopt.

The practical limitations to Friedrich's argument are significant, but the deeper problem lies with the moral principle that underpins it. The foundation of Friedrich's argument is the claim that we ought to prevent people from serious harm when we can do so at little cost to ourselves. If that principle is correct, one may wonder whether adoption is really the appropriate course of action to take. A recent estimate by the United States Department of Agriculture concludes that a child born in the United States in 2015 will cost $233,610 to raise (Lino et al. 2017). The problem for Friedrich's position is that *a lot* more harm will be prevented by donating that money (or even just a significant portion of it) to cost-effective charities, such as the Against Malaria Foundation or Schistosomiasis Control Initiative.[14] Thus, one may wonder why – at least for those who do not view children as an indispensable part of their life plans – this principle of preventing harm does not entail that they should refrain from procreating altogether: they could use the money saved to prevent a much greater quantity of harm through donating to cost-effective charity organizations. This alternative argument against procreation has been developed at length by Stuart Rachels (2014).

The duty to aid those in need

Rachels draws significantly on Peter Singer's (1972, 2009) work on world hunger and poverty. Singer (1972) defends the following moral principle: "If it is in our power to prevent something bad from happening, without thereby sacrificing anything of comparable moral importance, then we ought, morally, to do it" (p. 231). He also observes that there are millions of people across the world who are suffering and dying because they do not have adequate food, water, shelter or medical care. Suffering and death caused by a lack of food, water, shelter, and medical care are undeniably bad, and many of us have the ability to prevent these harms from occurring by donating to cost-effective charities. Moreover, these donations will often not deprive us of anything important. Almost anyone living in the western world surely purchases some luxuries that they could forego without any meaningful impact on her welfare. Thus, Singer reasons that many of us are morally obligated to donate a significant portion of our income to charities that will help prevent these harms.

Rachels views his argument as a variant of Singer's that is designed to illuminate one of its surprising implications: taking this duty to reduce suffering seriously requires that many of us refrain from procreating. Friedrich (2013) tries to avoid endorsing this position by arguing that we have a duty to prevent suffering only when we can do so "at little cost to ourselves" (pp. 25–26). He assumes that foregoing parenthood (both of biological and non-biological children) would be very costly for most people but also holds that foregoing parenthood only of one's biological children will often not be as costly. While this position is perfectly coherent, if one properly appreciates the moral weight of millions of people suffering and dying annually from easily preventable circumstances, it becomes more difficult to maintain. A desire to raise a child, in terms of moral significance, does not remotely compare to the suffering and death that could be prevented through $200,000 worth of donations to cost-effective charities. Can those living in developed nations really justify spending so much on their own children when they could save the lives of *many* other children who are on the brink of death elsewhere in the world?[15]

Rachels' argument is powerful, and while it would be psychologically challenging for most of us to live up to the standard that the argument requires – that is, to prioritize the prevention of suffering to such an extreme degree – such difficulty does not obviously remove the obligation. Even if we knew that we could not perfectly live up to that obligation, we could still have a duty to strive for it, and we might well come close to meeting the standard if we really tried. Even so, many feel that there must be a limit to what morality can reasonably demand of us and that this kind of obligation surpasses the threshold of what morality can require. The relevant question is why: if we cannot offer a good explanation, then this thought amounts to little more than complaining that being moral is difficult.

I discussed morality's demandingness to some extent in the previous chapter, but given its salience with respect to this argument, some further remarks are necessary. The limits of morality's demandingness originate from at least two sources: our limitations (both physical and psychological) as human beings and our desire to have a flourishing human life. When morality requires that we do something that we are physically or psychologically incapable of doing, then our limitations release us from a duty to perform the task. We cannot be obligated to do the impossible. The other limitation on morality's demands can be understood as follows: when a moral imperative proves antithetical to one's goal of living well, then sometimes that moral imperative should no longer be regarded as a strict obligation. I say *sometimes* because our life plans are often malleable enough to accommodate moral imperatives without undercutting our goal of living well. Without this qualification, the claim could serve to justify moral apathy in cases where it is inexcusable.

Aside from circumstances where women are prevented from taking action to prevent pregnancy or end pregnancy before birth, there is nothing physically impossible about refraining from procreation. The more common challenge to a duty to refrain from procreation is a mental one: not everyone may be capable of resisting the psychological urge to procreate. Rachels (2014) acknowledges that it may be permissible for such people to have children:

> I don't think it makes sense, either as social policy or as abstract philosophy, to hold people accountable for choices that are psychologically forced on them (even if they could physically do otherwise). For that reason, even though it would be regrettable for such people to have children (because their $227,000 could be better spent), I would not regard their decision to have children as immoral. Indeed, I'm not even sure I would regard it as a decision.
>
> (p. 578)

This exception is reasonable, but it likely does not apply to most people. Many of the motivations that people have for wanting children do not involve a desperate psychological need for them. It might be a preference, but a mere preference is not sufficient to ground the claim that one is psychologically *incapable* of doing otherwise.

If an imperative to procreate is objectionable because it is too demanding, then the problem will usually be that it would impede the goal of living well. Bernard Williams (1973), in his critique of utilitarianism, connects this limitation to an agent's integrity. He argues that utilitarianism is objectionable because any commitment an agent has must be abandoned as soon as it becomes inconsistent with maximizing utility. Williams (1973) imagines a person being required to abandon central life projects to fulfill the obligation to maximize utility and argues that it would be absurd to think this person is required to step aside from his own life project to adhere to the utilitarian determination of what leads to the best possible outcome. Because such a process alienates a person from their own actions and the connection between their actions and convictions, he views utilitarianism as an attack on a person's integrity (pp. 116–117). Williams might be exaggerating a bit, but the underlying point remains strong. We often structure our lives around certain personal and professional pursuits, and abandoning them would seem inconsistent with our character and opposed to our long-term life goals.

When moral imperatives force us to abandon the individual pursuits that serve as the central source of our lives' meaningfulness, sometimes those imperatives ought to be rejected.[16]

We must recognize that having children is an extremely strong desire for many people. For those who genuinely view rearing biological children as a central aspect of their life plans, it is unreasonable to demand that they remain childless, even if the money required to raise their children would do more good in the world if spent elsewhere.[17] However, if we are being honest, few people meet this criterion. How many people truly think so carefully and reflectively about the likely effects of their having a first (or second or third) child? Some procreate because they have grown bored with their current circumstances as a couple or because it is what others expect. Others just procreate because they view it as a normal part of life – the next big step in the lives of many young adults. People who procreate for these reasons are acting unjustifiably. To justify investing so much money into something when it could do so much good if put elsewhere, the investment must be contributing significantly to an important part of a person's life plans.[18] Thus, for those living in parts of the world where raising children is an expensive endeavor, procreating for trivial reasons is morally wrong.

A cautious approach to procreation

In this chapter, I have responded to a variety of different arguments defending antinatalism – the view that procreation is bad and (often) morally wrong. None of these arguments succeeded in justifying a wholesale condemnation of procreation, but they do convey an important moral lesson: the standards for justifying procreation are *much higher* than most people believe. There are powerful moral reasons that count *against* having children. Procreation risks serious harm to the child who is born, and the harms of life are imposed without the child's consent. Moreover, procreation involves serious financial costs in many parts of the world, and by spending that money to create a new life and support that life, a parent loses the opportunity to use that money to help the many already existing children who would benefit from it tremendously. People often procreate without taking these reasons into account. In these cases, the decision to procreate will usually lack moral justification, given the risk of harm to the child born and the other considerations in favor of remaining childless. Hence, a significant portion of people – particularly those who do not view procreation as a vital part of their life plans and who have significant financial resources at their disposal – should refrain from procreation and use some of the time and money saved to help others in dire need.

This outlook on procreation runs counter to the common beliefs that procreation is almost always permissible and something that should typically be praised, but the moral considerations that point us in this direction cannot be ignored. Those considerations help to explain why other authors who

126 *Approaches to procreative ethics*

reject antinatalism often reach the conclusion that many acts of procreation are unjustified and that the process of creating new human beings should be undertaken cautiously (e.g., Overall 2012; Wasserman 2015; Weinberg 2016). The practice of human reproduction is fraught with moral peril, and as a result, we must conclude that ordinary moral beliefs about the ethics of procreation are mistaken.

Notably, none of the antinatalist arguments examined in this chapter appeal to population growth to ground the impermissibility of procreation. When we add the environmental impacts of population growth to moral considerations that count in favor of antinatalism, the case in favor of a cautious approach to procreation becomes even stronger. While I contend that antinatalists are wrong to conclude that procreation is always wrong, they are right to subject our procreative norms to critical scrutiny and to demand serious moral justifications from those who procreate. Moreover, since procreation is morally problematic far more often than most people believe, we have an additional reason for pursuing the policy measures I mentioned in Chapter 5 that would encourage greater critical reflection on procreative decisions and result in a lower number of births.[19]

Notes

1 For an overview of some other ways in which antinatalism has appeared outside academia, see Tuhus-Dubrow (2019).
2 To Benatar's credit, he does attempt to address many of his critics. See Benatar (2013).
3 For some examples of this research, see Matlin and Stang (1978), Taylor (1989), Weinstein (1980, 1984), and Taylor and Brown (1998).
4 For more on the role that optimism and hope play in our species' survival, see Tiger (1979).
5 This taxonomy of what makes a person's life go well comes from Parfit (1987, pp. 493–502).
6 Benatar and his critics also usually neglect the additional complication of how difficult it may be to actually make sensible comparisons between the good and bad features of life. As Marsh (2014) notes, some of these goods and bads may be incommensurable with one another (p. 449, fn 23). Two items are incommensurable when the items are not equal to one another and neither of them are greater than (or less than) the other. If some of life's good and bad features are incommensurable, then there may not be a fact of the matter about whether certain lives are good or bad, all things considered. For an overview of value incommensurability, see Hsieh (2016).
7 Harrison and Tanner (2011) defend a similar argument, but they describe procreating as taking "an unjustifiable gamble that future generations will behave responsibly" (p. 114). Since they believe that human beings to do generally behave in morally responsible ways, they consider this gamble irrational and unjustified. Thus, they claim that we should not bring more humans into existence: "Human beings are dangerous things; too dangerous" (p. 114).
8 This claim does not imply that racism, sexism, and other forms of discrimination have been purged from society. The point is that equal rights and social standing

for people is gradually becoming the norm, which is a stark contrast to the dominant and overt racism and sexism of the past.
9 I owe this point to David Wasserman (2015, pp. 169–170).
10 In his own critique of Häyry, Wasserman (2015) also argues that the second condition is not met (pp. 175–176), but he grants (without explanation) that the first condition *does* seem to apply to the case of procreation. This is an odd concession. While we may not be able to assign exact probabilities, we have fairly strong empirical evidence that the vast majority of people regard their lives as worth living, and so I am not skeptical about our ability to make probabilistic estimates, perhaps using ranges (e.g., 90–95 percent), to estimate the likelihood that a person's life will be worth living.
11 Friedrich (2013) acknowledges that many people do not want to raise children and that raising them is extremely demanding (p. 28). Hence, he concedes that those who do not wish to have children can be exempt from the duty to adopt of those grounds. Thus, his argument applies only to those who intend to raise children.
12 Rulli (2016) identifies the desire to experience pregnancy as the most plausible exception to the general duty to adopt.
13 The surplus of parents interested in adoption has been generated in part because restrictions on international adoption have increased significantly in the last 15 years. As a result, international adoption rates have fallen dramatically. For an overview of this problem, see Montgomery and Powell (2018).
14 These charities often appear on GiveWell's list of most cost-effective charities. See GiveWell (2019) for their most recent rankings.
15 The World Health Organization (2016) estimates that 5.9 million children under the age of 5 died in 2015 and that over half these deaths could have been prevented with access to cheap, routine medical interventions.
16 It is important here to recognize that our life plans are not just about making ourselves happy. In the previous chapter, I mentioned that procreation often decreases parents' happiness rather than increasing it. That fact is fully compatible with parenthood being a rationally considered part of a person's life plan. Many parents might reason that decreases in happiness are outweighed by gains in other areas – perhaps an increase in meaningfulness or the cultivation of virtues like familial love and parental pride. Living well means more than just living happily.
17 Rieder (2015) makes a similar point in his discussion of the "gestational project" that women often view as a central life project (pp. 301–302).
18 This same reasoning could justify significant financial investments in other pursuits, such as learning how to play an instrument or engaging in expensive athletic training, so long as these pursuits are genuinely a significant part of a person's life plans.
19 Significant portions of this chapter are derived from chapter 3 of my doctoral dissertation. See Hedberg (2017).

References

Baumeister, Roy, Kathleen Vohs, Jennifer Aaker, and Emily Garbinsky. 2013. "Some Key Differences Between a Happy Life and a Meaningful One." *The Journal of Positive Psychology* 8, no. 6: 505–516.

Bayne, Tim. 2010. "In Defence of Genethical Parity." In *Procreation and Parenthood*, eds. David Archard and David Benatar, pp. 31–56. Oxford: Oxford University Press.

Belshaw, Christopher. 2007. Review of *Better Never to Have Been: The Harm of Coming Into Existence*, by David Benatar. *Notre Dame Philosophical Reviews*. http://ndpr.nd.edu/news/25313/?id=9983. Accessed November 16, 2019.

Benatar, David. 2006. *Better Never to Have Been*. Oxford: Clarendon.

Benatar, David. 2013. "Still Better Never to Have Been: A Reply to (More of) My Critics." *The Journal of Ethics* 17, no. 1: 121–151.

Benatar, David. 2015. "Part I: Anti-Natalism." In *Debating Procreation*, eds. David Benatar and David Wasserman, pp. 11–132. New York: Oxford University Press.

Bradley, Ben. 2010. "Benatar and the Logic of Betterness." *Journal of Ethics & Social Philosophy* 4, no. 2: 1–5. www.jesp.org/articles/download/BenataronBetternessNote.pdf. Accessed February 18, 2020.

Brown, Campbell. 2011. "Better Never to Have Been Believed: Benatar on the Harm of Existence." *Economics and Philosophy* 27, no. 1: 45–52.

Brickman, Philip, Dan Coates, and Ronnie Janoff-Bulman. 1978. "Lottery Winners and Accident Victims: Is Happiness Relative?" *Journal of Personality and Social Psychology* 36, no. 8: 917–927.

Brown, Jonathon, and Keith Dutton. 1995. "Truth and Consequences: The Costs and Benefits of Accurate Self-Knowledge." *Personality and Social Psychology Bulletin* 21, no. 12: 1288–1296.

Cantwell, Nigel. 2003. "Intercountry Adoption. A Comment on the Number of 'Adoptable' Children and the Number of Persons Seeking to Adopt Internationally." *The Judges' Newsletter on International Child Protection* 5. www.iss-ssi.org/2007/Resource_Centre/Tronc_DI/documents/CantwellIntercountryAdoptionENG.pdf. Accessed November 30, 2019.

DeGrazia, David. 2010. "Is It Wrong to Impose the Harms of Human Life? A Reply to Benatar." *Theoretical Medicine and Bioethics* 31, no. 4: 317–331.

Finer, Lawrence, and Mia Zolna. 2016. "Declines in Unintended Pregnancy in the United States, 2008–2011." *New England Journal of Medicine* 374, no. 9: 843–852.

Friedrich, Daniel. 2013. "A Duty to Adopt?" *Journal of Applied Philosophy* 30, no. 1: 25–39.

GiveWell. 2019. "Top Charities." www.givewell.org/charities/top-charities. Accessed December 12, 2019.

Graff, E. J. 2009. "The Lie We Love." *Foreign Policy*. http://foreignpolicy.com/2009/10/06/the-lie-we-love/. Accessed November 30, 2019.

Harman, Elizabeth. 2009. "David Benatar. Better Never To Have Been: The Harm of Coming into Existence." *Nous* 43, no. 4: 776–785.

Harrison, Gerald, and Julia Tanner. 2011. "Better Not to Have Children." *Think* 10, no. 27: 113–121.

Häyry, Matti. 2004. "A Rational Cure for Prereproductive Stress Syndrome." *Journal of Medical Ethics* 30: 377–378.

Hedberg, Trevor. 2017. "Population, Consumption, and Procreation: Ethical Implications for Humanity's Future." Ph.D. dissertation, Department of Philosophy, University of Tennessee.

Hsieh, Nien-hê. 2016. "Incommensurable Values." *Stanford Encyclopedia of Philosophy*. http://plato.stanford.edu/entries/value-incommensurable/. Accessed November 30, 2019.

Kaposy, Chris. 2009. "Coming Into Existence: The Good, the Bad, and the Indifferent." *Human Studies* 32, no. 1: 101–108.

Lammerant, Isabelle, and Marlène Hofstetter. 2007. *Adoption: At What Cost? For an Ethical Responsibility of Receiving Countries in Intercountry Adoption*. Geneva, Switzerland: Terre des Hommes International Federation.

Licon, Jimmy. 2012. "The Immorality of Procreation." *Think* 11, no. 32: 85–91.

Lino, Mark, Kevin Kuczynski, Nestor Rodriguez, and TusaRebecca Schap. 2017. *Expenditures on Children by Families, 2015*. U.S. Department of Agriculture, Center for Nutrition Policy and Promotion. https://fns-prod.azureedge.net/sites/default/files/crc2015_March2017_0.pdf. Accessed November 30, 2019.

Marsh, Jason. 2014. "Quality of Life Assessments, Cognitive Reliability, and Procreative Responsibility." *Philosophy and Phenomenological Research* 89, no. 2: 436–466.

Matlin, Margaret, and David Stang. 1978. *The Polyanna Principle: Selectivity in Language, Memory, and Thought*. Cambridge, MA: Schenkman.

Montgomery, Mark, and Irene Powell. 2018. *Saving International Adoption: An Argument from Economics and Personal Experience*. Nashville, TN: Vanderbilt University Press.

Nelson, Jamie. 2016. Review of *Permissible Progeny? The Morality of Procreation and Parenting*, by Sarah Hannan, Samantha Brennan, and Richard Vernon (eds.). *Notre Dame Philosophical Reviews*. https://ndpr.nd.edu/news/permissible-progeny-the-morality-of-procreation-and-parenting/. Accessed November 16, 2019.

Overall, Christine. 2012. *Why Have Children?* Cambridge, MA: MIT Press.

Pandey, Geeta. 2019. "Indian Man to Sue Parents for Giving Birth to Him." *BBC News, Delhi*. www.bbc.com/news/world-asia-india-47154287. Accessed November 16, 2019.

Parfit, Derek. 1987. *Reasons and Persons*. Oxford: Oxford University Press.

Pinker, Steven. 2011. *The Better Angels of Our Nature: The Decline of Violence in History and its Causes*. New York: Viking.

Rachels, Stuart. 2014. "The Immorality of Having Children." *Ethical Theory and Moral Practice* 17, no. 3: 567–582.

Rawls, John. 1999. *A Theory of Justice: Revised Edition*. Cambridge, MA: Belknap Press.

Rieder, Travis. 2015. "Procreation, Adoption, and the Contours of Obligation." *Journal of Applied Philosophy* 32, no. 3: 293–309.

Rulli, Tina. 2016. "Preferring a Genetically-Related Child." *Jorunal of Moral Philosophy* 13, no. 6: 669–698.

Sedgh, Gilda, Susheela Singh, and Rubina Hussain. 2014. "Intended and Unintended Pregnancies Worldwide in 2012 and Recent Trends." *Studies in Family Planning* 45, no. 3: 301–314.

Shiffrin, Seana. 1999. "Wrongful Life, Procreative Possibility, and the Significance of Harm." *Legal Theory* 5: 117–148.

Singer, Peter. 1972. "Famine, Affluence, and Morality." *Philosophy and Public Affairs* 1, no. 3: 229–243.

Singer, Peter. 2009. *The Life You Can Save*. New York: Random House.

Smuts, Aaron. 2014. "To Be or Never to Have Been: Anti-Natalism and a Life Worth Living." *Ethical Theory and Moral Practice* 17, no. 4: 711–729.

Taylor, Shelley. 1989. *Positive Illusions: Creative Self-Deception and the Healthy Mind*. New York: Basic Books.

Taylor, Shelley, and Jonathon Brown. 1998. "Illusion and Well-Being: A Social Psychological Perspective on Mental Health." *Psychological Bulletin* 103, no. 2: 193–210.

Tiger, Lionel. 1979. *Optimism: The Biology of Hope*. New York: Simon and Schuster.

Tuhus-Dubrow, Rebecca. 2019. "I Wish I'd Never Been Born: The Rise of the Anti-natalists." *Guardian*. www.theguardian.com/world/2019/nov/14/anti-natalists-childfree-population-climate-change. Accessed November 16, 2019.

UNICEF. 2018. "Orphans." www.unicef.org/media/orphans. Accessed December 12, 2019.

Wasserman, David. 2015. "Part II: Pro-Natalism." In *Debating Procreation*, eds. David Benatar and David Wasserman, pp. 133–264. New York: Oxford University Press.

Weinberg, Rivka. 2016. *The Risk of a Lifetime: How, When, and Why Procreation May Be Permissible*. New York: Oxford University Press.

Weinstein, Neil. 1980. "Unrealistic Optimism about Future Life Events." *Journal of Personality and Social Psychology* 39, no. 5: 806–820.

Weinstein, Neil. 1984. "Why It Won't Happen to Me: Perceptions of Risk and Susceptibility." *Health Psychology* 3, no. 5: 431–457.

Wellman, Christopher Heath. 2008. "Immigration and Freedom of Association." *Ethics* 119, no. 1: 109–141.

Williams, Bernard. 1973. "A Critique of Utilitarianism." In *Utilitarianism: For and Against*, eds. J. J. C. Smart and Bernard Williams, 77–150. Cambridge: Cambridge University Press.

Wood, Joanne. 1996. "What is Social Comparison and How Should We Study It?" *Personality and Social Psychology Bulletin* 22, no. 5: 520–537.

World Health Organization. 2016. "Children: Reducing Mortality." www.who.int/mediacentre/factsheets/fs178/en/. Accessed November 30, 2019.

8 Reproductive rights and procreative freedom

With antinatalism addressed, I now turn to a different objection to my view. One of the lingering worries about taking action to reduce population growth is that doing so will infringe on people's reproductive rights and their reproductive freedom more generally. Many regard procreative liberty as extremely valuable. After all, procreation is often central to a person's identity, a source of meaningfulness, and an expression of a couple's love for one another. Since it often plays a crucial role in people's lives, John Robertson (1994) states that our ethical outlook should "recognize a presumption in favor of most personal reproductive choices" (p. 24). The importance of reproductive freedom is also expressed via its inclusion in the Universal Declaration of Human Rights as the right to found a family (UN 1948).

Given the importance of being able to govern one's own procreative acts, we need to consider whether any of the positions I have sketched in prior chapters would violate people's reproductive rights or otherwise infringe on their procreative liberty in objectionable ways. As has already been acknowledged, coercive population policies of the past often violated people's rights, sometimes in the form of forced abortion or sterilization. Nothing I have endorsed would approach that level of coercion or invasiveness. Moreover, as I mentioned in Chapter 5, outright coercive population policies have significant shortcomings even if all invasive measures are completely avoided.

I have endorsed autonomy-enhancing measures such as pursuing gender justice, improving sex education, and increasing access to contraception. These measures increase procreative autonomy either by increasing the range of choices that people have or by making their choices better informed. These measures do not threaten procreative freedom; they enhance it. Thus, those who champion reproductive freedom should fully support these proposals.

Of the methods for responding to population growth that I endorsed, only the semi-coercive measures could be seen as a threat to procreative freedom. I support preference-adjusting interventions and possibly some forms of incentivization. (As I mentioned in Chapter 5, I am wary of using incentive-based schemes, but I think a blanket prohibition on them would be too hasty.) So the central question is whether these specific strategies for lowering fertility would constitute an unwarranted infringement on procreative freedom.

Is the right to procreate unlimited?

We can start with a quick appraisal of a rather extreme view – the position that reproductive rights entail an unlimited (or unrestricted) right to procreate. On this view, any infringement on reproductive rights, no matter how small, is considered morally unacceptable.[1] Procreation certainly is one of people's most fundamental interests, but is such an extreme position plausible? I argue it is not because no rights – not even the most fundamental ones – are limitless in scope.

Everyone has the right to life, but we recognize that we may permissibly violate this right when, for instance, a person threatens the lives of innocent others. Everyone has a right to bodily autonomy, but we recognize that this right does not permit me to use my arms and fists to assault someone else. Rights can be constrained when exercising them inflicts harm on other people or conflicts with the rights of those other people.[2] Procreation can clearly harm other people. As highlighted in the previous chapter, procreating carries a risk of severe harm to the child who is born. Procreation can also harm already existing children. If the birth of a third child leads the two existing children to be neglected or leaves the parents unable to support their original two children, then that act of procreation caused harm to those children. Finally, procreation can contribute to environmental impacts that harm both present and future people. After all, this new person will be a participant in society and in all likelihood engage in many of the same environmentally destructive behaviors that we do. These considerations provide grounds for recognizing limits on the right to procreate.

Among recent authors who have written on this subject, Sara Conly (2016) makes perhaps the most thorough case that the right to procreate cannot be unlimited. She even suggests that the right to have a child should be interpreted as the right to have only *one* child, at least under our current circumstances. According to her reasoning, rights must correspond to fundamental interests that we have, and the fundamental interests that are fulfilled by procreating can be adequately fulfilled by having just one child. In some cases, the interests associated with parenting can be met through adopting a child and thereby not require procreation at all. Even if someone has a specific interest in having a biological child, that interest can be fulfilled by having a single child. One child is enough to continue genetic lineage, after all. If we have an interest in having a family, that interest can also be fulfilled by having a single child. Having a *large* family might require additional procreation, but the right to procreate does not entail having a right to "the family that fulfills one's dreams" (Conly 2016, p. 51). We do not have a right to a family with four children any more than we have a right to a child who will become a professional athlete. Additionally, the notion that a larger family is better than a smaller family is dubious, and the central goods associated with family life do not require the participation of several biological children. Adoption also presents an option for increasing family size without procreating.

The right to procreate could also track an interest in being regarded by others as equally worthy of reproducing, but this desire for equal standing could be achieved under many circumstances that involve greatly restricted procreative freedom. If everyone were only permitted to have one child, then everyone would have equal standing with respect to procreation so long as this constraint was applied to everyone equally. In less extreme scenarios, equal standing can be achieved so long as the social norms and expectations regarding procreation apply equally to all involved. So if a social norm to have families of two or fewer children develops, that does not threaten someone's sense of equal standing unless there are some people who are exempt from this social expectation.[3]

The right to procreate without restriction might also be derived from a broader right to bodily autonomy, but that strategy faces an immediate obstacle. The right to bodily autonomy only allows us to exercise this freedom when doing so does not harm others, and procreation can harm others – both the person born and others affected by that person's use of resources. On these grounds, appealing to a right of bodily autonomy will not be sufficient to justify procreation in cases where it causes harm.

To counter this point, one might argue that procreative acts do not cause harm in the way relevant to limiting our rights. Consider Travis Rieder's (2016) remarks in his review of Conly's *One Child*:

> Conly is of course correct that having a right to bodily autonomy doesn't mean that one can do whatever she likes with her body. Although I have a right to swing my arm, I do not have a right to swing my arm where your face is located. However, this sort of argument is problematic in the context of overpopulation … *my procreating doesn't harm anyone* through its contribution to overpopulation. Precisely as she notes, environmental problems like climate change make traditional moral reasoning hard, because they involve massively complex collective action, and it just doesn't seem true that my taking almost any single action harms anyone. In a population of 7.3 billion people, any number of people that I can add to the population makes *virtually no difference* – the resources consumed by my child, against the earth's available resources, are *infinitesimal*.
> (pp. 30–31, original emphasis)

Rieder does not think that procreative acts cause harm in the way that would justify restricting a person's rights because one more child born has such a miniscule impact on the Earth in the grand scheme of things. This line of reasoning should sound familiar: it is one of the objections that can be raised to the claim that an individual's greenhouse gas emissions cause harm. Conly does not offer a substantive response to this concern,[4] and positing an account of harm that vindicates her judgment in this case would be challenging. Nonetheless, I believe there is an argument that can vindicate Conly's view that the right to bodily autonomy does not grant a right to unrestricted procreation.

One plausible constraint on fundamental rights is that the exercise of these rights cannot be incompatible with respecting the fundamental rights of other people. In other words, fundamental rights impose a type of constraint on other fundamental rights. In a very basic case, a person's right to personal security places a constraint on my right to bodily autonomy: except in unusual circumstances, I cannot assault another person. The case with population is more complicated, but the underlying principle is no different. Millions of people will suffer and die as a result of climate change, biodiversity loss, and other forms of severe environmental degradation if population growth is not curtailed. The collective exercise of an unlimited right to procreate will cause the rights of others – both in the present and future – to be violated, and the rights violated will be among the most critical rights they have: the right to life, the right to health, and the right to means of subsistence (Caney 2010; Shue 2011).[5] These are rights that must be fulfilled just to ensure a person's physical survival – the most basic rights a person can have (Shue 1996). A right to procreate is important, but it is not *that* important. No one dies from a failure to procreate; people do die if their basic rights go unfulfilled. So these other rights should take priority over the right to procreate when they come into conflict. Thus, the right to procreate cannot be understood as a right to *unlimited* procreation: it is limited by the extent to which its collective exercise affects the ability of others to have their basic rights respected.

In the past, there was often no danger of undermining others' rights by procreating excessively: in fact, for the vast majority of human history, we needed to be rather prolific in our procreation just to ensure the continuation of our species. But our circumstances have changed, and so our limits on the right to procreate must change as well. Does this mean that our right to procreate should be understood as allowing us to have only one child under present conditions? Conly (2016) believes that it does (pp. 217–220), and given the stakes, it is hard to argue. The right to procreate should be understood only as guaranteeing a right to have one child, at least so long as the global population size contributes so significantly to our environmental problems.[6] As I argued in Chapter 6, some couples might be permitted to have two children, but the onus is on them to provide a powerful justifying reason for having a second biological child. I do not think most couples will be able to meet that standard, so most prospective parents will need to limit themselves to one child to fulfill their procreative moral obligations.

At this juncture, I must reiterate that I do not endorse any strict coercive measures to induce compliance to this obligation. To recall a point from Chapter 5, these measures compromise reproductive autonomy severely without being clearly superior to other non-coercive measures, and proposing them may only serve to further discourage people from openly discussing population issues again. So I am not advocating that we abandon procreative freedom in favor of draconian population reduction schemes. The central message of this section is that the right to procreation is not unlimited and that minor infringements on reproductive freedom are not automatically objectionable rights violations. This observation is pivotal because some semi-coercive measures can reduce people's procreative autonomy by changing social norms or creating new incentives to engage in certain behaviors. This fact alone does not mean that they are automatically objectionable, though, because some minor reductions to procreative freedom can be justified under our current circumstances.

Semi-coercive measures and reproductive freedom

The fact that semi-coercive measures sometimes reduce reproductive freedom does not give us sufficient reason to reject these measures, but it does give us reason to examine them a little more carefully. Could these methods compromise reproductive autonomy enough that we think they should not be pursued? Rebecca Kukla (2016) provides some reasons for thinking that this might be the case.

One of Kukla's concerns is that pursuing preference-adjustment strategies through mass media would not be able to avoid sending the message that smaller families are a more responsible choice than larger families. Should these views become entrenched in the background culture, then women may no longer feel like they have the "unburdened option of choosing a larger family" (Kukla 2016, p. 872). She also worries that this could result in a loss of social and economic support for large families. Access to prenatal testing created a prevailing norm of discontinuing pregnancies when the fetus has a genetic defect, and disability advocates worry that this norm could lead to reduced resources for those who choose to carry a disabled child to term (Kukla 2016, p. 873). Kukla has a similar fear about parents who want large families under conditions where antinatalist values become widespread. She also highlights the ways in which vulnerable groups (e.g., the economically disadvantaged, women of color, the disabled) could be particularly harmed by these new cultural norms and the ways in which large-family stigma would hinder the reproductive autonomy of women even in the developed world.

As an initial response, we might question how plausible it is that antinatalist values become sufficiently entrenched for these problems to materialize. Pronatalist values have a very deep foothold in most societies, including in those where fertility has dipped below replacement levels. It is still the norm to celebrate announcements of pregnancy and to regard childrearing as an expectation of growing older.[7] In many cultures, a life without parenthood is seen as suboptimal or outright regrettable. It would be rather remarkable if attitudes toward procreation shifted from positive to negative under these conditions. Preference-adjusting interventions may just make people more open-minded to the acceptability of small families.

An alternative response is to say that these outcomes, even if they could happen, are avoidable. Hickey, Rieder, and Earl (2016) claim that targeted messaging and careful forethought regarding the tactics used can avoid the concerns Kukla has in mind (p. 861). Perhaps that is true for some of them, but it sounds too optimistic to suppose *all* the undesirable effects are avoidable. In any case, there are better replies available.

The most straightforward response is to deny that social norms toward smaller families would be a bad thing overall, even if not all of the negative effects could be avoided. One of the key takeaways from previous chapters is that usually having a small family is morally better than having a large family. It may sound odd when put so bluntly, but it is a direct implication of the arguments put forward in Chapter 6. People who limit their procreation to (at most) two children are adhering to a moral obligation that others are violating. Now given the personal and intimate nature of procreation, it would be both callous and inappropriate to use shame- or guilt-based tactics to motivate people to have smaller families. Simultaneously, we want people to engage in morally responsible behavior, and so we want our social norms to encourage morally responsible behavior. Thus, norms that promote reflection on procreative decisions and a cautious approach to conceiving a child seem like improvements to the status quo.

To further elaborate on this point, recall the discussion of antinatalism in the previous chapter. While I did not find any of the arguments supporting antinatalism to be wholly successful, the moral considerations underpinning them indicate that the standards for permissible procreation are very high. Many acts of procreation are undertaken without the appropriate level of caution or critical reflection on one's motivations for procreation. As a result, some children born under these circumstances are subjected to an unacceptable risk of harm. Since these acts are impermissible, changing cultural norms to favor a lower rate of procreation would be preferable to the status quo. As things stand, procreative acts are too often approached without the care and deliberation that they warrant. Thus, a shift toward antinatalism seems preferable *independent of any concerns about population*. The fact that we also need to reduce population just makes the case in favor of antinatalist preference adjustment even more compelling.

Additionally, while it is important not to overlook the fact that some people may be made worse off by a change in cultural norms regarding procreation, we must likewise not overlook the fact that there are currently people made worse off by cultural norms that put pressure on them to procreate. As Eileen Crist (2019) notes, "some of the grossest violations of human rights are perpetrated in societies that force women to start having children when they are barely beyond childhood themselves, and to continue reproducing until their bodies are no longer fertile" (p. 207). Moreover, many societies employ incentive-based schemes or even legal penalties (e.g., by criminalizing abortion) to encourage citizens to have more children. Thus, for large numbers of people, alleviating the pressure to procreate would boost their autonomy and make it easier for them to live as they wish.[8] Neither the status quo nor a small-family social norm will be perfect, but a shift in the direction of smaller families would be a huge improvement overall.

Escaping the specter of population control

One of the main reasons that people are so protective of reproductive freedom is that past policy measures that aimed at lowering fertility often involved exercising control over women's bodies. The phrase "population control" has since come to be associated with unsavory practices designed to rob women of their reproductive autonomy. As Joni Seager (1993) states, "Population control is a euphemism for the control of women" (p. 216). Similar sentiments are expressed by Betsy Hartmann (1987) when she argues that the population problem has been exaggerated to justify violating women's rights. Some authors have also argued that efforts at population control, rather than being concerned for the global poor or the environment, were actually just eugenics in disguise – deliberate attempts to control which social groups reproduce and which ones do not (Hardt and Negri 2004, pp. 165–166; Wilson 2012). The unsavory history tied to population control initiatives casts a long shadow. As a result, any mention of efforts to change fertility rates carry the fear that such proposals are not really motivated by a concern for environmental sustainability or the welfare of future people but by more insidious agendas, such as reinforcing patriarchal values or limiting the procreation of racial minorities.[9]

We should not forget the mistakes of the past, but we can also not let them paralyze us. We are now more than 25 years past the United Nations International Conference on Population and Development – the venue where explicit discussion of population policy became a political taboo. Evading the problem has not helped us. Population growth has continued and made it more difficult to mitigate climate change, slow down the rate of species extinctions, and adequately distribute the world's finite resources. Minimizing the harm that befalls present and future people requires confronting this reality and abandoning the fiction that procreative choices are too private or intimate to be subjected to moral scrutiny.

The good news from a practical standpoint is that the objectives of promoting reproductive freedom and reducing population size overlap considerably. Our best means of halting population growth will not involve fining people obscene sums of money for procreating above some government-specified limit or subjecting women to mandatory abortions. But they should involve raising awareness of the connection between procreation and our environmental problems, and they should try to normalize small families. Depending on how much progress we can make with these measures, we may need to examine some incentivization programs to further encourage morally responsible behavior – perhaps changing the tax structure to give benefits to those with fewer dependents rather than more or financial compensation for attending family planning classes. These measures would reduce the procreative freedom of some people, but they need not constitute a violation of their reproductive rights and would not be morally objectionable. In morally tragic circumstances, it is not possible to ensure a perfectly just outcome for everyone. In these instances, the welfare of the present and future people whose livelihoods are at stake must take priority over the reproductive freedom of others.

Notes

1 See Abrams (1996) for a defense of this position.
2 If I exercise my right to control my body assaulting another person then I am both causing them harm and violating their right to personal security.
3 Even then, social norms are often not regarded as unduly coercive, and they are unlikely to constitute a serious threat to reproductive freedom unless it is accompanied by significant penalties for noncompliance.
4 She acknowledges that the picture of harm is more complicated in the case of climate change than more typical circumstances (Conly 2016, p. 93), but as Rieder (2016, p. 31) says, she does not seem to properly appreciate the significance of this observation.
5 Here, I am assuming that it is appropriate to regard future people as having the same rights as present people. The rationale for this assumption was described in defense of the *Equity of Non-Harm* principle in Chapter 3. Future people, at least those who will exist in the next few centuries, will have the same fundamental interests as present people. Since rights protect fundamental interests, future people should have the same rights as present people. While future people do not yet exist, we can still do things now that lead to their rights being violated when they do exist. Such actions are just as wrong as violating the corresponding rights of present people.
6 For further discussion of the right to procreate, see Conly (2016, ch. 2–3), Kates (2004), and Overall (2012, ch. 2).
7 Even with generally favorable attitudes toward procreation, there is some significant international variance in the availability and affordability of childcare services. If antinatalist social norms became more pronounced in countries where these services are already inadequate, then we can envision ways in which parents would be disadvantaged by that result. The pivotal question is whether the preference adjustment would take place at such a scale where this result manifest.
8 For some additional thoughts on this point, see Hickey, Rieder, and Earl (2016, pp. 859–861).
9 For further discussion of this concern, see Coole (2018, pp. 68–74).

References

Abrams, Paula. 1996. "Reservations about Women: Population Policy and Reproductive Rights." *Cornell International Law Journal* 29, no. 1: 1–41.
Caney, Simon. 2010. "Climate Change, Human Rights, and Moral Thresholds." In *Climate Ethics: Essential Readings*, edited by Stephen Gardiner, et al., pp. 163–177. New York: Oxford University Press.

Conly, Sarah. 2016. *One Child: Do We Have a Right to Have More?* Oxford: Oxford University Press.

Coole, Diana. 2018. *Should We Control World Population?* Cambridge: Polity Press.

Crist, Eileen. 2019. *Abundant Earth: Toward an Ecological Civilization.* Chicago: University of Chicago Press.

Hardt, Michael, and Antonio Negri. 2004. *Multitude: War and Democracy in the Age of Empire.* New York: Penguin Books.

Hartmann, Betsy. 1987. *Reproductive Rights and Wrongs: The Global Politics of Population Control and Contraceptive Choice.* New York: Harper & Row.

Hickey, Colin, Travis Rieder, and Jake Earl. 2016. "Population Engineering and the Fight against Climate Change." *Social Theory and Practice* 42, no. 4: 845–870.

Kates, Carol. 2004. "Reproductive Liberty and Overpopulation." *Environmental Values* 13, no. 1: 51–79.

Kukla, Rebecca. 2016. "Whose Job Is It to Fight Climate Change? A Response to Hickey, Rieder, and Earl." *Social Theory and Practice* 42, no. 4: 871–878.

Overall, Christine. 2012. *Why Have Children?* Cambridge, MA: MIT Press.

Rieder, Travis. 2016. "Review: Sarah Conly, One Child: Do We Have a Right to Have More?" *Kennedy Institute of Ethics Journal* 26, no. 2: 29–34.

Robertson, John. 1994. *Children of Choice: Freedom and the New Reproductive Technologies.* Princeton, NJ: Princeton University Press.

Seager, Joni. 1993. *Earth Follies: Coming to Feminist Terms with the Global Environmental Crisis.* New York: Routledge.

Shue, Henry. 1996. *Basic Rights: Subsistence, Affluence, and U.S. Foreign Policy*, 2nd ed. Princeton, NJ: Princeton University Press.

Shue, Henry. 2011. "Human Rights, Climate Change, and the Trillionth Ton." In *The Ethics of Global Climate Change*, ed. Denis Arnold, 292–314. Cambridge: Cambridge University Press.

UN (United Nations). 1948. *Universal Declaration of Human Rights.* www.un.org/en/universal-declaration-human-rights/. Accessed December 10, 2019.

Wilson, Kalpana. 2012. *Race, Racism, and Development: Interrogating History, Discourse, and Practice.* London: Zed Books.

Part IV
Lingering questions

9 What about immigration?

The previous chapters in this book outline an argument that we have a moral duty to reduce our population size, assess what policy measures could permissibly be pursued to achieve that goal, evaluate what specific procreative obligations individuals have, and address a variety of objections to those positions. In this final part of the book, I briefly address three lingering questions related to the content of previous chapters. The first of these concerns what implications my discussion of population growth may have for immigration policy.

The moral and political issues that surround immigration are a vexing morass of competing concerns and conflicting principles. We recognize the unjust circumstances that lead people to want to relocate, but we also recognize that groups and organizations usually have the right to place restrictions on their membership. We recognize the general right of freedom of movement but also recognize the importance of respecting a nation's sovereignty. Amidst these conflicts (and others) exist a list of difficult questions about the political power of the state, the equality of people regardless of nationality, the commitments of liberal democratic societies, the relevance of historical injustice, and the value of preserving extant cultures.[1] As expected, it is beyond my means to resolve this debate here. Instead, I will focus on the two main ways that population-related considerations are relevant to the ongoing immigration debate.

First, I will argue that population-related considerations provide a significant moral reason for nations with large ecological footprints to *restrict* immigration, at least under certain circumstances.[2] Second, I will argue that in nations where population reduction is occurring at a swift pace, *increasing* immigration could serve as a permissible means of easing the population shrinkage and making the transition to a smaller population more gradual. While neither of these arguments will resolve the larger immigration debate, they should provide some insight as to how population-based considerations could play a role in determining what immigration policy is the most just.

The argument from moral cosmopolitanism, intergenerational version

The argument from moral cosmopolitanism is one of the most intuitively compelling arguments for an open-border immigration policy – a policy in which people are free to cross national borders and relocate to another country with few or very limited restrictions.[3] This argument begins with the empirical observation that geographical location plays a huge role in one's life prospects. This observation is then paired with the moral claim that all human beings have equal moral worth.[4] In practice, affirming this belief entails regarding all people as having equal moral value and being equally worthy of moral protection except in extreme cases (e.g., when a person has committed a heinous crime). If we endorse this moral outlook, then we must wonder what could justify preventing people from relocating to improve their life prospects. If a person born in Kenya is just as morally valuable as someone born in the United States, then what justifies the people of the United States preventing Kenyan citizens from relocating to the United States to improve their life prospects? One's place of birth is beyond one's control, and thus, limiting a person's life prospects just because they were born in a particular country seems deeply unjust. If we are to do what justice demands, then we ought to have open borders and allow people to relocate as they please.

While this argument may not be decisive, most would acknowledge that it carries significant moral weight and offers at least one good reason to favor open borders.[5] But when this argument is given an intergenerational interpretation, it actually might favor a closed border position. Moral cosmopolitanism can be understood in two different ways:

- Current-Generation Moral Cosmopolitanism (CGMC): all presently existing human beings have equal moral worth.
- Intergenerational Moral Cosmopolitanism (IMC): all human beings have equal moral worth regardless of when they exist.

Based on arguments that I have made in previous chapters, it should not be surprising that I favor IMC. With respect to their moral standing, when someone exists is just as morally irrelevant as where they exist. Thus, moral cosmopolitanism should be extended to apply to future people. Doing so has interesting implications for what constitutes a just immigration policy.

If we are to minimize the harm that befalls future people, we must work to address climate change, biodiversity loss, resource depletion, unsustainable agricultural practices, and various other forms of environmental degradation that threaten their well-being. Most developed nations have massive ecological footprints, and they will need to reduce these footprints in response to these problems. If the populations of these nations continue to rise, then it will be more difficult for them to fulfill their duties to reduce their ecological footprints because the required reductions in their environmentally harmful consumption will be steeper. Thus, if most developed nations are going to reduce their ecological footprints to the extent required, they will need to maintain a stable population or, in some cases, ensure that their population gradually shrinks.

Given that most developed nations have fertility rates below replacement, this might not seem like a problem, but some nations are still seeing their domestic populations increase due to immigration. When immigrants relocate from a developing country to a developed country, their ecological footprints increase significantly.[6] As a result, it becomes more difficult for that country to reduce its overall environmental impact. This fact can be offered as a reason to restrict immigration into developed nations. Philip Cafaro and Winthrop Staples III (2009) offer an argument of this sort focused on the immigration policy of the United States:

1 Immigration levels are at a historic high and immigration is now the main driver of U.S. population growth.
2 Population growth contributes significantly to a host of environmental problems within our borders.
3 A growing population increases America's large environmental footprint beyond our borders and our disproportionate role in stressing global environmental systems.
4 To seriously address environmental problems at home and become good global environmental citizens, we must stop U.S. population growth.
5 We are morally obligated to address our environmental problems and become good environmental citizens.
6 Therefore, we should limit immigration into the U.S. to the extent needed to stop U.S. population growth.

(pp. 5–6)[7]

While there is no explicit appeal to IMC in this argument, something similar to it is likely lurking in the background. The reason to be so concerned with solving these environmental problems must be at least partially tied concerns about the welfare of future people and the recognition that their needs are being compromised to provide benefits to present people. The imperative to take the suffering and death of future people seriously lends some significant moral weight to the claim that nations like the U.S. should restrict immigration.

Nevertheless, these kinds of arguments for immigration restriction have not persuaded everyone. One objection is that focusing on a single nation's immigration policy is an ineffective strategy for addressing environmental problems that are global in scope (Crist 2019, p. 209; Nagtzaam 2018, p. 718). An adequate response to overpopulation must be carried out with a significant degree of international coordination. One nation focusing on its own immigration policy will not make much of a difference.

As a criticism of an argument that focuses only on the United States, this is a reasonable criticism, but the argument could easily be broadened to apply to a wider range of countries. Suppose the argument looked like this instead:

1★ Presently, a host of environmental problems severely threaten the welfare of both present and future people.
2★ To adequately respond to these environmental problems, developed nations must reduce their ecological footprints rapidly and significantly.
3★ Developed nations will not be able to reduce their environmental footprints rapidly and significantly if their domestic populations grow.
4★ Therefore, developed nations should limit immigration into their countries to the extent needed to stop domestic population growth.

The conclusion expressed in (4★) could, if it were followed, make a significant difference in helping developed nations to curtail their environmentally harmful consumption and reduce their impacts of the environment. But that does not mean that the argument is sound.

A more pressing concern is that these kinds of restrictions on immigration are unjust. Eileen Crist (2019) argues that the United States, Canada, Australia, and Western Europe should not be allowed to "exclude entry to people whose lives those rich nations have actively and knowingly contributed to destroying through a self-serving commodity regime, a fossil-fuel based economy, and policies of dumping their toxic wastes in developing nations" (p. 209). There is little doubt that wealthy nations have contributed to the problems that plague poorer nations in various ways, so this is a forceful response to the claim that wealthy nations should be allowed to restrict immigration to help lower their own ecological footprints. But is it decisive?

Crist's concerns reveal a way in which the immigration issue may be morally tragic. In Chapter 5, I introduced this idea to describe scenarios in which no available course of action can prevent injustice from occurring. Immigration may present us with some scenarios that are moral tragedies. If future people are morally equal to present people, then protecting their welfare from the threats posed by ongoing environmental problems like climate change is a very high moral priority: hundreds of millions, if not billions, of people may be adversely affected by these environmental impacts. There are undoubtedly many people who have been adversely impacted by developed countries – people who have legitimate claims of justice to be allowed to relocate to those nations. But future people also have legitimate claims to have their most fundamental interests protected. If allowing widespread immigration to developed nations makes the long-term effects on future people worse, then future people are being treated unjustly.

Climate change in particular presents a concrete form of this dilemma. The greenhouse gas emissions of developed nations have contributed the most significantly to climate change, so there is a strong case for claiming that they are morally obligated to accommodate those in developing nations who are displaced by it (Nawrotzki 2014). However, admitting climate refugees will increase the population of these developed nations and thereby increase their overall ecological footprint and make it harder for them to lower their greenhouse gas emissions. Limiting immigration to only *climate* refugees might lessen the severity of this problem, but the cost of doing so is that other refugees, whose lives are in immediate danger, would need to be turned away. What should we do in these kinds of scenarios?

I suggested in Chapter 5 that the best general strategy for addressing moral tragedies is to pick the option that minimizes the injustice that occurs, but in this case, it would be very difficult to determine the numbers of people affected or the severity of the injustice the relevant parties are experiencing. An additional complication is that domestic population growth can be lowered in alternative ways (as discussed in Chapter 5), so we would have to consider their effectiveness in accomplishing that goal relative to a restriction on immigration policy. The good news is that there may be a solution to this conundrum that does not require us to resolve these issues.

Immigration, economics, and population stabilization

One concern about population reduction is that it will reduce economic growth. I responded briefly to one version of this objection in Chapter 5, emphasizing that a fixation on economic growth should be abandoned in a world where resources are limited. There is, however, a better version of this objection that connects a thriving economy to duties of justice. Population reduction is typically associated with economic stagnation or even outright recession (Prettner, Bloom, and Strulik 2013). If fertility drops below replacement levels, then the proportion of older members of society increases gradually over time. This results in economic stagnation because there are fewer working class members of society and can lead to lower standards of living if the government has to make cuts to manage the nation's budget. The bigger problem, however, is that this economic stagnation could make it harder for a nation to fund the mitigation and adaptation measures needed to fight climate change or diminish its ability to contribute to international anti-poverty efforts (Earl, Rieder, and Hickey 2017, p. 584). Poverty remains a serious global problem, and a decreased ability to respond to it could affect hundreds of millions of people. In 2015, the World Bank (2019) estimates that 736 million people were living in extreme poverty, which means they had an income of less than $1.90 per day.

Countries with fertility rates of 1.5 face a serious possibility of economic stagnation or recession if nothing changes.[8] In the short term, such costs could be quite significant. One option in this situation would be to try to raise fertility back up to the replacement rate, but it would be counterproductive to the aim of reducing population size in the long term to raise fertility rates now. Fortunately, immigration provides a potential solution to this problem.

Developed nations that are well below replacement fertility could promote immigration and reduce their restrictions on who can legally enter their country. Since young immigrants from developing nations already have some incentive to relocate to developed nations, simply making immigration easier can serve to offset the population reduction taking place due to low fertility. This strategy would also enable these countries to accommodate a greater number of refugees, helping them fulfill some duties of justice to present people. Since immigrants also tend to have lower ecological footprints than native-born citizens of developed nations, replacing these native-born citizens with immigrants results in an ecological net gain (though not as big as one in which no immigration occurs).[9]

The picture sketched above might be the best overall solution that can be reached at the intersection of immigration and population. Fertility rates would be well below replacement, so population reduction in the nation would occur. But to ensure that the population does not shrink so quickly as to create significant economic difficulties in the short term, a greater number of immigrants can be admitted annually. This strategy makes progress toward the long-term goal of population reduction while also alleviating the short-term burdens on present people. It also better fulfills duties of justice toward refugees and other immigrants who have been harmed by the actions of developed nations than an extremely restrictive immigration policy would.

The critical observation to make about this solution is that it is only possible if the fertility rate in the country well below replacement. In the United States, for instance, fertility levels are at about 1.9 children per woman (CIA 2019), but because of its annual immigration, its population continues to grow. If the fertility rate in the United States were lower, then it could allow for a similar number of immigrants, or perhaps even more, without growing its population. It may not always be possible to get fertility rates so low, but it is worth pursuing because it appears to enable to reasonable compromise between the moral needs of present and future people alike.

With this in mind, these considerations are far from solving the bigger puzzle of what a just immigration policy would be. What I hope to have done in this short chapter is highlight two considerations – the relationship between immigration and domestic population growth and the role of immigrations in alleviating economic challenges associated with rapid domestic population decrease – that should play a role in the ongoing discussions about immigration. Environmentalists have generally retreated from discussing immigration and its connection to environmental impacts (Cafaro 2015, ch. 7), but these two considerations demonstrate that there are areas where they could make meaningful contributions to this conversation. I hope they will speak up a bit more in the future.

Notes

1 Even as broad as these considerations are, they still do not accurately capture the complexity of the immigration debate. For an overview, see Wellman (2019).
2 Of course, a significant reason to restrict immigration is not necessarily a decisive one. I will leave it open whether this reason could be overridden by other moral considerations.
3 For some presentations of this argument, see Cole (2012) and Carens (1987, 2013, ch. 11).
4 This is the core tenet underlying moral cosmopolitanism. See, e.g., Miller (2016, p. 22).
5 This explains why defenders of closed-border positions generally make significant efforts to respond to this objection. For some examples, see Blake (2013), Miller (2016, ch. 2), and Wellman (2008).
6 Immigrants to developed countries may have a slightly lower ecological footprint than those born in developed countries. As one example, immigrants to the United States have carbon footprints 18 percent smaller than those of native-born Americans (Camarota and Kolankiewicz 2008). Even so, it is clear that their relocation entails a significant increase in their ecological footprint compared to if they remained in a developing nation.
7 For another presentation of this argument, see Cafaro (2015, chs. 7 and 8).
8 Some research suggests that declining fertility is actually compatible with continued growth, albeit at a slower rate, in developed countries (Quamrul, Weil, and Wilde 2013) and that fertility reduction can actually advance growth significantly in developing countries (Lee, Mason, and members of the NTA Network 2014). However, I will concede the empirical basis of the objection for the sake of argument.
9 For further discussion of this solution to the short-term economic obstacles created by population reduction, see Earl, Rieder, and Hickey (2017, pp. 585–586).

References

Blake, Michael. 2013. "Immigration, Jurisdiction, and Exclusion." *Philosophy & Public Affairs* 41, no. 2: 103–130.

Cafaro, Philip. 2015. *How Many Is Too Many? The Progressive Argument for Reducing Immigration into the United States*. Chicago: University of Chicago Press.

Cafaro, Phil, and Winthrop Staples III. 2009. "The Environmental Argument for Reducing Immigration into the United States." *Environmental Ethics* 31, no. 1: 5–30.

Camarota, Steven, and Leon Kolankiewicz. 2008. "Immigration to the United States and World-Wide Greenhouse Gas Emissions." Center for Immigration Studies. https://cis.org/Immigration-United-States-and-WorldWide-Greenhouse-Gas-Emissions-0. Accessed December 17, 2019.

Carens, Joseph. 1987. "Aliens and Citizens: The Case for Open Borders." *Review of Politics* 49, no. 2: 251–273.

Carens, Joseph. 2013. *The Ethics of Immigration*. Oxford: Oxford University Press.

CIA (Central Intelligence Agency). 2019. "Country Comparison: Total Fertility Rate." *The World Factbook*. www.cia.gov/library/publications/the-world-factbook/rankorder/2127rank.html. Accessed December 7, 2019.

Cole, Phillip. 2012. "Taking Moral Equality Seriously: Egalitarianism and Immigration Controls." *Journal of International Political Theory* 8, nos. 1–2: 121–134.

Crist, Eileen. 2019. *Abundant Earth: Toward an Ecological Civilization*. Chicago: University of Chicago Press.

Earl, Jake, Colin Hickey, and Travis Rieder. 2017. "Fertility, Immigration, and the Fight against Climate Change." *Bioethics* 31, no. 8: 582–589.

Lee, Ronald, Andrew Mason, and members of the NTA Network. 2014. "Is Low Fertility Really a Problem? Population Aging, Dependence, and Consumption." *Science* 346, no. 6206: 229–234.

Miller, David. 2016. *Strangers in Our Midst: The Political Philosophy of Immigration*. Cambridge, MA: Harvard University Press.

Naagtzaam, Gerry. 2018. Review of How Many is Too Many? The Progressive Argument for Reducing Immigration into the United States. *Environmental Values* 27, no. 6: 716–718.

Nawrotzki, Raphael. 2014. "Climate Migration and Moral Responsibility." *Ethics, Policy & Environment* 17, no. 1: 69–87.

Prettner, Klaus, David Bloom, and Holger Strulik. 2013. "Declining Fertility and Economic Well-Being: Do Education and Health Ride to the Rescue." *Labour Economics* 22, no. C: 70–79.

Quamrul, H. Ashraf, David N. Weil, and Joshua Wilde. 2013. "The Effect of Fertility Reduction on Economic Growth." *Population Development Review* 39, no. 1: 97–130.

Wellman, Christopher Heath. 2008. "Immigration and Freedom of Association." Ethics 119, no. 1: 109–141.

Wellman, Christopher Heath. 2019. "Immigration." *Stanford Encyclopedia of Philosophy*. https://plato.stanford.edu/entries/immigration/. Accessed November 8, 2019.

Wilcox, Shelley. 2009. "The Open Borders Debate on Immigration." *Philosophy Compass* 4, no. 5: 813–821.

World Bank. 2019. "Poverty: Overview." www.worldbank.org/en/topic/poverty/overview. Accessed December 18, 2019.

10 What about the nonhuman community?

Up to this point, I have approached overpopulation and its environmental impacts focusing solely on human interests and values. This approach is anthropocentric: it assumes that only human beings have direct moral standing and that animals and the environment only matter insofar as those things matter to human beings.

The Population Reduction Argument that I defended in Chapters 3 and 4 makes no reference to animals or nonhuman entities to ground a moral duty to reduce our population size. To the extent that considerations like biodiversity loss or environmental degradation were examined, they were only done so to the extent that they are valuable to human beings. I adopted this approach because I wanted to show that there were significant moral reasons to stabilize and reduce the human population even if we limit our scope to the concerns of humanity (as we often do in a variety of other contexts). Even so, I do not endorse an exclusively anthropocentric approach to ethics, and in this chapter, I want to relax the anthropocentric assumption and consider the implications for my position if we grant some members of the nonhuman community direct moral standing.

If conscious animals matter

On July 7, 2012 at the first annual Francis Crick Memorial Conference, a team of neuroscientists drafted and publicly signed *The Cambridge Declaration on Consciousness* (Low et al. 2012). In doing so, they gave authoritative voice to an observation that plenty of people, both scientists and laypeople, had already made: many animals beyond human beings are conscious creatures capable of joy and suffering. The scientific literature on the cognitive capacities of nonhuman animals had converged on this conclusion long before this declaration (Proctor, Carder, and Cornish 2013), and anyone with a pet dog or cat probably needed no convincing of this claim. Still, an official declaration from experts on the subject matters because our treatment of animals so often neglects the reality that they are conscious and can feel pain and pleasure just as we do.

The scientists who signed *The Cambridge Declaration on Consciousness* did so because they were familiar with scientific studies involving the biology, physiology, and behavior of animals. Many animals, especially mammals and birds, have similar neural circuitry to us, and so from a neuroscientific standpoint, it is unsurprising that such animals can feel pain. Yet the same capacity is also found in octopi – an invertebrate that has a much different nervous system than mammals or birds. We are still learning about animals' capabilities and the cognitive capacities, but it is obvious that many of them have similar or greater intelligence than some groups of human beings. As adults, the smartest animals – apes, dolphins, and large dogs (to name only a few) – clearly have greater cognitive abilities than infants. Yet we accord infants an extremely high level of moral protection, usually on a par with the protections we extend to adult human beings. Why then do we not do the same with these adult animals? Perhaps the answer is that an infant will have a much higher cognitive capacity than these animals when it is fully grown. But that is not always true. Some humans with severe cognitive impairments will never have the same intelligence, autonomy, or communication skills as those possessed by the smartest nonhuman animals.[1] We would never condone treating these people the way that we often treat nonhuman animals, but we cannot appeal to a difference in cognitive capacity to explain this asymmetry in treatment. Thus, we are left to wonder whether it actually *can* be justified.

Animal consciousness raises a powerful challenge to methods of moral theorizing that focus solely on human beings. Conscious experience entails the ability to experience suffering and enjoyment. Conscious beings have at least one fundamental interest – the promotion of pleasure rather than pain. This is what morally separates a conscious being from a mere object. A brick or fencepost has no subjective experience. They cannot feel pain or pleasure and cannot exercise any sort of conscious agency to interact with the world around them. Thus, whether these objects are dropped, thrown, cut, or disintegrated does not set back any of their interests, and so what happens to these objects makes no moral difference (unless these objects are used in ways that affect the welfare of other conscious beings). These objects do not have any welfare, so nothing can go better or worse for them. Conscious beings, in contrast, do have welfare, and it can be raised or lowered by our actions. Playing fetch with an energetic puppy will raise its welfare; breaking its leg will lower its welfare. Since our actions can affect animals' welfare, we must take their interests into account when we make moral decisions.

The moral significance of animal consciousness has led many ethicists to attempt to incorporate it into their own moral theorizing (e.g., DeGrazia 1996; Engster 2006; Korsgaard 2018; Midgley 1983; Palmer 2010; Regan 1983; Sapontzis 1987; Singer 2002; Varner 2012; Wood 1998). The trend has been so widespread across different theoretical approaches to ethics that the dominant question about animals is no longer "Do animals matter morally?" but rather "How can animals be incorporated into our moral thinking?" The emerging consensus on this issue entails that a moral theory limited only to the interests and values of human beings is, at best, incomplete.

Unfortunately, many of our ways of treating animals still reflect an extraordinary bias in favor of our own species. Some are tempted by the thought that human beings are unique in some way that justifies according them a higher moral status than all other animals, including conscious ones. As mentioned earlier, the phenomenon of severe cognitive impairment raises major challenges for this view, but even human beings generally do have a *higher* moral status than nonhuman animals, it does not follow that animals have *no* moral status or that the difference between humans and animals is so great that we are justified in ignoring animals' interests. To revisit an important point from Chapter 3, one of the most basic moral imperatives is the duty to avoid causing unnecessary harm to someone else. Conscious animals, since they can experience pain and have their interests stifled, can be harmed. Thus, at a bare minimum, our treatment of conscious animals should not cause them unnecessary harm.

Even using only the basic moral imperative not to harm animals unnecessarily, we can identify a staggering amount of wrongdoing that is taking place. Every year, more than 70 billion land animals are slaughtered for human consumption (UN Food and Agricultural Organization 2019).[2] This is almost ten times the total number of human beings who inhabit the Earth, and this level of killing takes place *every year*. Most of these animals are raised in industrial settings, often called "factory farms," where they suffer severely before being slaughtered. The objective in factory farms is to produce as much meat as efficiently as possible. The motivation for this approach is entirely economic: having more animal products to sell leads to higher profits. Since animals are viewed purely as economic resources in these settings, their welfare is not a high priority. The result is that animals suffer greatly so that meat can be cheap and widely available.[3]

Factory farms are more efficient when more animals are contained within their walls, so the animals typically live in cramped and crowded conditions. In this environment, animals often have to be harmed to prevent them from injuring one another. Chickens are de-beaked so that they cannot peck one another while packed so tightly together. Pigs will often bite or chew on one another's tails, so their tails are usually clipped to eliminate that possibility. To accelerate their growth, the animals are often injected with hormones. Sometimes, chickens will grow so quickly that their legs lack the strength to support their body weight. As a result, they must remain sitting for the remainder of their short lives. The animals in these facilities are consistently exposed to their own feces and vomit, so they are at high risk for infections and other ailments. Even heavy doses of antibiotics are often not enough to prevent the animals from debilitating sickness. Overall, factory-farmed animals live brief lives dominated by pain and discomfort, usually in cramped environments where they are unable to exercise their natural capacities.[4]

The suffering imposed on factory-farmed animals is severe yet produces only a minor benefit to people – the availability of cheap meat. Can so much suffering and death really be justified by such a meager benefit to humans? Given what we know about animal consciousness and its connection to having interests, the answer is surely no.[5] If we give the welfare of animals any meaningful weight, then treating animals in this manner is morally impermissible. Yet tens of billions of them suffer such a fate every year. And that number does not even include the creatures that are harvested from the ocean. If we were to add them to the tally of animals killed by humans annually, the total number might well surpass one trillion.[6]

Because of the sheer quantity of animals that suffer and die for human consumption, animal agriculture is one of the gravest moral problems in the world. This remains true even if we think that animal suffering and death are far less morally significant than the suffering and death of human beings. Suppose that we judged the unnecessary death of a conscious animal to be only 0.1 percent as morally significant as the death of an adult human being. That would mean that we would have to kill 1000 animals for it to have the same moral significance as killing just one adult human being. But there are over 70 billion conscious animal deaths annually, so even at this ratio, our practices in animal agriculture would still be as morally bad as something that caused more than 70 million human deaths per year. To put that figure in perspective, 58.4 million human beings died in all of 2019 (UN Department of Economic and Social Affairs 2019, p. 13).

As outlandishly high as the number of animal deaths already is, this figure will almost surely increase in the near future. Research from the UN Food and Agricultural Organization projects that demand for meat will rise 73 percent between 2010 and 2050 (Gerber et al. 2013). This estimate accords with the worldwide trend toward increased meat production. Over 70 billion animals were slaughtered for human consumption in 2017, but less than 50 billion were slaughtered for our consumption in 2000 (FAO 2019).

Meat demand is rising for two reasons. First, when nations develop, their populations tend to increase their meat consumption. Several highly populated nations, including China and India, are expected to see a huge rise in meat demand as their economic circumstances improve. Second, as human population grows, there are more mouths to feed. Since meat eating is such a common human behavior, more people generally results in more meat consumption. In tandem, these two factors make it inevitable that meat demand will rise significantly as the century progresses.

Many animal activists advocate a societal shift toward vegetarian or vegan diets, but the global trend is pulling in the opposite direction. A widespread abandonment of meat consumption will not happen anytime soon. Now that does not mean that trying to get people to eat less meat is pointless: raising awareness of the morally unacceptable aspects of current farming practices and advocating for social change are critical aspects of working toward a long term solution. My point is that trillions of animals will suffer and die before that solution materializes. So until we shift to an agricultural system that imposes far less harm on animals,[7] we must consider alternative ways to reduce animal suffering. It is obvious that the size of our population will play a huge role in the overall demand for meat, so one clear strategy for reducing animal suffering is to pursue population reduction.

This would also help to reduce animal suffering in other ways. As our population grows, we must expand and encroach on territory inhabited by wild animals. Even when people do not have ill intentions toward animals, they still have basic needs that must be met, and when nonhuman species must compete with human beings for essential space and resources, those nonhuman species usually lose (Mckee 2003). Habitat destruction and fragmentation leaves many animals unable to find a suitable environment in which to live, and unsurprisingly some of them die. As already discussed, our numbers play a huge role in global climate change, and rising global temperatures do not just affect humans. Some species are unable to relocate but find it much harder to survive in the higher temperatures. Others relocate to higher elevations or to different regions but struggle to integrate themselves into a new ecological niche.

All told, our world features an extraordinary amount of animal suffering and death, and much of it is caused directly by human actions. Conscious animals have direct moral standing, and the vast majority of the animal suffering and death that we cause is unnecessary. Population reduction could play a significant role in decreasing the amount of animal suffering and death that occurs in the future. Thus, if we properly acknowledge the moral significance of animal consciousness, we have even stronger moral reasons to adopt policy measures that lower fertility rates than if we focus entirely on human interests and values.

If all living things matter

Some environmental ethicists do not think that it is only conscious animals that matter. Biocentrism is the view that all living things have direct moral standing, including nonconscious living things like trees and bacteria.[8] This outlook stems from the belief that these organisms, just by virtue of being alive, have goods of their own. Essentially, what this means is that things can go better or worse for them. Even if an organism is not conscious, things can happen to it that either impede or promote its biological functioning. Plants fare better when they have the proper amount of sunlight and water; they fare worse when they do not have these things. We can refer to the measure of an organisms biological functioning as its biotic welfare (Nolt 2015, pp. 176–179).

The claim that biotic welfare is morally significant is much more controversial than the claim that consciousness is morally significant. Not everyone is convinced that merely being alive confers direct moral standing. After all, a non-conscious life form cannot experience pain or pleasure no matter how good or bad its biological functioning. Chopping down a tree does not cause that tree pain – it lacks the biological structures necessary to have that sensation. It also does not stifle any of the trees' desires or preferences, since one must be conscious to form desires or preferences. Thus, even if we acknowledge that the tree has a biotic welfare, we may not think that this particular type of welfare is morally significant.

Even if we assume that biotic welfare is valuable, there are questions about how its value should be quantified. We might hold that there is an infinite difference in value between a conscious being's welfare and the welfare of a merely living being.[9] That would entail that no number of merely living beings would have the moral significance of even one conscious being. So, as an example, no amount of bacteria could be as morally significant as a dog. If we adopted this perspective, then it would technically be a version of biocentrism, but since we would never prioritize merely living beings over conscious beings, it would be virtually indistinguishable from an ethic that only regards conscious beings as having direct moral standing.

While I am somewhat skeptical that moral standing should extend beyond conscious beings, I will not explore the matter further here. Instead, I simply want to highlight that biocentrists have *even stronger* reasons for supporting population reduction than those who think the bounds of moral standing stop at consciousness. If biotic welfare carries significant moral weight, then the specter of the Earth's sixth mass extinction event now carries a far stronger and grimmer moral weight (Barnosky et al. 2011). Species extinctions are now not just a loss of beauty, knowledge, ecosystem services for humans or a source of suffering and death for conscious animals. For all beings (including nonconscious ones), extinction represents a permanent disappearance of a source of biotic welfare.[10] Even in scenarios where species do not go extinct, the destruction of the natural world now carries a much greater moral significance. Consider the destruction of coral reefs via ocean acidification or the deforestation taking place in the Amazon. The sheer number of living organisms adversely affected by these activities translates to a staggering loss of biotic welfare.

The central point is that if you regard biotic welfare as morally significant, then you have even more reason to believe we should seek to reduce population. For those who think all living things have moral standing, our ongoing problems will appear far more severe than those who view them only through an anthropocentric lens or with an eye only toward the interests of conscious beings. This same trend will hold true for those who want to extent moral standing even further, such as to species as a whole (Rolston 1988) or to ecosystems (Leopold 1970). The further we extend moral standing into the natural world, the worse our circumstances look. And the worse our circumstances appear to be, the more urgent the need to reduce our population size becomes.

Nonanthropocentrism and human population

The massive size of the human population contributes to an enormous amount of animal suffering and to extraordinary losses of biotic welfare.

Giving even a shred of moral standing to conscious animals or other living things makes the case for population reduction much stronger, and the more weight we accord to the welfare losses tied to the nonhuman community, the more urgent our duty to reduce our population size becomes. For these reasons, those who advocate for the moral status of animals or the moral standing of the environment more broadly should be the loudest voices promoting population reduction.

I have argued throughout other areas of this book that we have a moral duty to pursue population reduction even if we limit our moral concern to the interests and values of human beings. However, there is compelling evidence that some entities other than humans have direct moral standing. The case that conscious animals have such standing is all but irrefutable in light of the empirical evidence about their cognitive capacities. A failure to account for the suffering of conscious animals shows either ignorance or an unjustifiable bias in favor of our own species. How far moral standing extends beyond conscious beings is much less certain, but even if we limit the scope of our moral concern to human beings and conscious animals, the hundreds of billions of animals that are killed each year to provide sustenance for us provide overwhelming evidence that our current practices are unjustifiable. One of the biggest reasons why we produce so much meat and harvest so many fish is to feed our burgeoning population. Reducing our numbers would mean fewer of these animals have to suffer and die. Halting our population growth would be a boon to the welfare of both future people and our nonhuman brethren.

Notes

1 For some examples involving specific animals and a discussion of the challenge that cognitive disability raises for our treatment of nonhuman animals, see Singer (2009).
2 The reported kill counts of animals across the world can be found by navigating the statistical database hosted by the Food and Agricultural Organization of the United Nations. According to this data, more than 74 billion land animals were slaughtered for food in 2017. Chickens (66.6 billion slaughtered), Ducks (three billion slaughtered), and pigs (1.4 billion slaughtered) account for more than 95 percent of the animals killed.
3 In this discussion, I focus solely on how industrialized animal agriculture leads directly to unnecessary animal suffering and death, but it is worth mentioning that it also contributes significantly to climate change, although the estimates of its contribution vary significantly. One report compiled by the UN Food and Agricultural Organization concludes that livestock production is responsible for roughly 18 percent of annual GHG emissions (Steinfeld et al. 2006). Other studies have estimated the contribution of animal agriculture to climate change at 14.5 percent of annual greenhouse gas emissions (Gerber et al. 2013) and between 19 percent and 29 percent of annual greenhouse gas emissions (Vermeulen, Campbell, and Ingram 2012).
4 The wretched and unsanitary conditions of factory farms are well documented. For a few representative examples, see Fitzgerald (2003), Pew Commission (2008), and Foer (2009).

5 Virtually all environmental ethicists agree on this point. Factory farming has been condemned from a wide range of moral traditions and on many different grounds. For some examples, see Anomaly (2015), Engster (2006), Francione and Charlton (2015), Hursthouse (2006), Regan (1983), and Singer (2002).
6 The number of marine animals that die annually is quite difficult to estimate due to variance in regional fisheries and the fact that catches are reported in weight rather than individual fish caught. Estimates range from 90 billion (ADAPTT 2019) to over one trillion (Mood and Brooke 2010).
7 All agricultural practices impose some suffering on animals. Even vegan systems do so through use of pesticides and the clearing of land for crops. So the goal of an agricultural system should be to minimize harm to animals, not to prevent it.
8 For some classic representations of biocentrism, see Naess (1973, 1987), Schwietzer (1923), and Taylor (1986).
9 See Hedberg (2017) for a brief argument in favor of this claim. For a response, see Nolt (2017).
10 There are efforts underway to develop means of resurrecting extinct species, but it is unclear at this time whether and to what extent they will be successful. For an overview, see Shapiro (2017).

References

ADAPTT (Animals Deserve Absolute Protection Today and Tomorrow). 2019. *More Than 150 Billion Animals Slaughtered Every Year.* www.adaptt.org/about/the-kill-counter.html. Accessed December 16, 2019.

Anomaly, Jonathan. 2015. "What's Wrong with Factory Farming?" *Public Health Ethics* 8, no. 3: 246–254.

Barnosky, Anthony, Nicholas Matzke, Susumu Tomiya, Guinevere Wogan, Brian Swartz, Tiago Quental, Charles Marshall, Jenny McGuire, Emily Lindsey, Kaitlin Maguire, Ben Mersey, and Elizabeth Ferrer. 2011. "Has the Earth's Sixth Mass Extinction Already Arrived?" *Nature* 471: 51–57.

Curnutt, Jordan. 1997. "A New Argument for Vegetarianism." *Journal of Social Philosophy* 28, no. 3: 153–172.

DeGrazia, David. 1996. *Taking Animals Seriously: Mental Life and Moral Status.* Cambridge: Cambridge University Press.

Engster, Daniel. 2006. "Care Ethics and Animal Welfare." *Journal of Social Philosophy* 37, no. 4: 521–536.

Fitzgerald, Deborah. 2003. *Every Farm a Factory: The Industrial Ideal in American Agriculture.* New Haven, CT: Yale University Press.

Foer, Jonathan. 2009. *Eating Animals.* New York: Little, Brown and Company.

Francione, Gary, and Anna Charlton. 2015. *Animal Rights: The Abolitionist Approach.* Exempla Press.

Gerber, P. J., H. Steinfeld, B. Henderson, A. Mottet, C. Opio, J. Dijkman, A. Falcucci, and G. Tempio. 2013. *Tackling Climate Change Through Livestock: A Global Assessment of Emissions and Mitigation Opportunities.* Food and Agricultural Organization of the United Nations. www.fao.org/docrep/018/i3437e/i3437e.pdf. Accessed December 16, 2019.

Gruen, Lori. 2017. "The Moral Status of Animals." *Stanford Encyclopedia of Philosophy.* https://plato.stanford.edu/entries/moral-animal/. Accessed December 16, 2019.

Hedberg, Trevor. 2017. "Review of Environmental Ethics for the Long Term: An Introduction." *Ethics, Policy & Environment* 20, no. 1: 121–124.

Hursthouse, Rosalind. 2006. "Applying Virtue Ethics to Our Treatment of Other Animals." In *The Practice of Virtue: Classic and Contemporary Readings*, edited by Jennifer Welchman, 136–155. Indianapolis, IN: Hackett.

Jamieson, Dale. "The Rights of Animals and the Demands of Nature." *Environmental Values* 17, no. 2: 181–200.

Korsgaard, Christine. 2018. *Fellow Creatures: Our Obligations to Other Animals*. Oxford: Oxford University Press.

Leopold, Aldo. 1970. *A Sand County Almanac with Essays from Round River*. New York: Ballantine books.

Low, Phillip, Jaak Panksepp, David Edelman, Bruno Van Swinderen, and Christof Koch. 2012. *The Cambridge Declaration on Consciousness*. http://fcmconference.org/img/CambridgeDeclarationOnConsciousness.pdf. Accessed December 16, 2019.

McKee, Jeffrey. 2009. *Sparing Nature: The Conflict Between Human Population Growth and Earth's Biodiversity*. New Brunswick, NJ: Rutgers University Press.

Midgley, Mary. 1983. *Animals and Why They Matter*. Athens, GA: University of Georgia Press.

Mood, Alison, and Phil Brooke. 2010. *Estimating the Number of Fish Caught in Global Fishing Each Year*. http://fishcount.org.uk/published/std/fishcountstudy.pdf. Accessed December 16, 2019.

Naess, Arne. 1973. "The Shallow and the Deep, Long-Range Ecology Movement. A Summary." *Inquiry* 16, no. 1–4: 95–100.

Naess, Arne. 1987. "Self-Realization: An Ecological Approach to Being in the World." *Environmental Ethics* 12, no. 2: 185–192.

Nolt, John. 2015. *Environmental Ethics for the Long Term: An Introduction*. New York: Routledge.

Nolt, John. 2017. "Are There Infinite Welfare Differences among Living Things?" *Environmental Values* 26, no. 1: 73–89.

Palmer, Clare. 2010. *Animal Ethics in Context*. New York: Columbia University Press.

The Pew Charitable Trusts. 2008. *Putting Meat on the Table: Industrial Farm Animal Production in America*. Baltimore: Pew Commission on Industrial Farm Animal Production. www.pewtrusts.org/en/research-and-analysis/reports/0001/01/01/putting-meat-on-the-table. Accessed December 16, 2019.

Proctor, Helen, Gemma Carder, and Amelia Cornish. 2013. "Searching for Animal Sentience: A Systematic Review of the Scientific Literature." *Animals (Basel)* 3, no. 3: 882–906.

Regan, Tom. 1983. *The Case for Animal Rights*. Berkeley, CA: University of California Press.

Rolston, Holmes, III. 1988. *Environmental Ethics: Duties to and Values in the Natural World*. Philadelphia, PA: Temple University Press.

Sapontzis, Steve. 1987. *Morals, Reasoning, and Animals*. Philadelphia, PA: Temple University Press.

Schwietzer, Albert. 1923. *Civilization and Ethics*, John Paull Naish (trans.). London: A. & C. Black.

Shapiro, Beth. 2017. "Pathways to De-extinction: How Close Can We Get to Resurrection of an Extinct Species?" *Functional Ecology* 31. no. 5: 996–1002.

Singer, Peter. 2002. *Animal Liberation*. New York: HarperCollins Publishers, Inc.

Singer, Peter. 2009. "Speciesism and Moral Status." *Metaphilosophy* 40, nos. 3–4: 567–581.

Steinfeld, H., P. Gerber, T. Wassenaar, V. Castel, M. Rosales, and C. De Haan. 2006. *Livestock's Long Shadow: Environmental Issues and Options*. Rome: Food and Agriculture Organization of the United Nations. www.fao.org/docrep/010/a0701e/a0701e.pdf. Accessed December 18, 2019.

Taylor, Paul. 1986. *Respect for Nature: A Theory of Environmental Ethics*. Princeton, NJ: Princeton University Press.

UN Food and Agricultural Organization. 2019. The Food and Agriculture Organization Corporate Statistical Database. www.fao.org/faostat/en/#home. Accessed December 8, 2019.

UN Department of Economic and Social Affairs. 2019. *World Mortality 2019: Data Booklet*. www.un.org/en/development/desa/population/publications/pdf/mortality/WMR2019/WorldMortality2019DataBooklet.pdf. Accessed December 18, 2019.

Varner, Gary. 2012. *Personhood, Ethics, and Animal Cognition: Situating Animals in Hare's Two-Level Utilitarianism*. Oxford: Oxford University Press.

Vermeulen, Sonja J., Bruce M. Campbell, and John S. I. Ingram. 2012. "Climate Change and Food Systems." *Annual Review of Environment and Resources* 37: 195–222

Wood, Allen. 1998. "Kant on Duties Regarding Nonrational Nature." *Aristotelian Society Supplementary Volume* 72: 189–210.

11 Can we solve the problem?

In the second chapter of this book, I considered whether concerns about population growth were overblown and argued that they were not. Now, as we near the end of our discussion, we must consider a concern that comes from the opposite end of the spectrum. One lingering worry about all these problems – climate change, species extinctions, resource depletion, freshwater shortages, and the like – is that we have already passed the point where we could avoid their most catastrophic effects. Even if we pursue the measures I have mentioned in this book and make extraordinary efforts to curb our rates of harmful consumption, will we be able to solve the problem in the time that remains?

The wrong question

Some people believe it is already too late for humanity. Consider these remarks from Guy McPherson (2016) regarding the rise in global temperature:

> The recent and near-future rises in temperature are occurring and will occur at least an order of magnitude faster than the worst of all prior Mass Extinctions. Habitat for human animals is disappearing throughout the world, and abrupt climate change has barely begun. In the near future, habitat for Homo sapiens will be gone. Shortly thereafter, all humans will die.

Since retiring from the University of Arizona in 2009, McPherson has devoted much of his time to publicizing his belief that humanity will soon go extinct. His efforts include a radio show (called *Nature Bats Last*), an active YouTube channel, various public lectures and workshops, and several written works.[1] In addition to defending his empirical claim about humanity's pending extinction, McPherson also offers advice about how to cope with this fact. Given the number of people who engage with his material, many people are clearly worried about the threat that climate change and related environmental problems pose for human civilization.[2]

Despite the following that McPherson has gained, the specific claim on which he grounds his views – namely, the prediction that all of humanity will go extinct within the next decade or two – is far too strong to endorse. It is important not to downplay the gravity of our current situation, but his diagnosis errs in the opposite direction. McPherson overestimates the severity and pace of environmental degradation and simultaneously underestimates human ingenuity and resilience. Humanity is not going extinct anytime soon. The real problem is that the future will contain significant losses for humans and nonhumans alike. This outcome cannot be prevented: after all, it is already happening. People are already suffering, species are already disappearing, and islands are already vanishing beneath the rising ocean. These observations can cause distress, and since individuals cannot change what is happening on their own, that distress can quickly give rise to apathy and despair.

Feelings of despondence and helplessness are understandable reactions to our current situation, but they should be resisted. These reactions are rooted in a misconception about our situation. To see that misconception, we must return to the question in the chapter title: can we solve the problem?

The question itself reveals an inaccurate way of understanding moral problems. It encourages us to see them as either solvable or unsolvable and to see a sharp divide between these categories. If a moral problem is solvable, then obviously we try to solve it. If a moral problem is unsolvable, then there's nothing to be done. After all, it would be pointless to attempt something that is impossible. According to this logic, if our environmental problems cannot be solved, then trying to solve them is a waste of time. Apathy becomes justified.

We can certainly recognize that some moral problems fit this framework. Moreover, some of them are clearly solvable. Suppose that your phone is stolen from you at a subway station, but the theft was captured on security cameras. A few days later, police apprehend the thief, your stolen phone is returned, and the thief is punished. In this case, it appears the moral problem – an impermissible theft – is resolved. We encounter plenty of other examples regularly as well. If we say something offensive at a social gathering, we can apologize and make amends. If we break a promise without good reason, we can acknowledge our mistake and offer some form of compensation. The small-scale moral problems of our day-to-day lives often have straightforward solutions.

Grand-scale moral problems do not typically work this way, though. They rarely allow for comprehensive solutions, and some of them might not even be "solvable" in any strict sense. Take racism as an example. The elimination of race-based discrimination is undeniably a worthy moral goal. But is it possible to *solve* the problem of racism – to eliminate it completely? I doubt it. Certain forms of racial discrimination are likely to persist no matter how much progress is made. As recent history has shown, even when racism is no longer socially acceptable and when race-based discrimination is rendered illegal, this form of bias nevertheless persists. In fact, even among people who consciously reject racism, racist attitudes and preferences can still manifest through subconscious cognitive processes.[3] All things considered, racism may be an unsolvable moral problem,[4] but that does not change the fact that eliminating racism is a worthy moral pursuit. The moral progress made in the pursuit of racial equality has resulted in more just social conditions and improved the welfare of various marginalized populations. These achievements are not diminished by the fact that we have not fully expunged racism from existence.

There is no doubt that our ongoing environmentally destructive behaviors will have grave consequences later this century. In some cases, recovering from the damages will be costly and difficult; in other cases, the damage may be irreversible. But these facts do not render the pursuit of environmental sustainability any less valuable. The scope of our impacts will vary dramatically depending on what actions we take in the remainder of this century. Some harmful impacts have already come to pass, and many more will come. But how many? And how prepared will we be for them? We can still affect the answers to these questions, and so long as that is true, we should not be gripped by the paralysis of apathy or despair.[5]

We cannot *solve* problems like climate change or biodiversity loss: it is not possible to shield present and future people from the harms on the horizon, and it is not possible to prevent all vulnerable species from going extinct. We can, however, reduce the severity of these phenomena and make them more manageable for ourselves and our descendants. It would be very bad for the world to warm 3 °C above pre-industrial global temperatures, but a temperature rise of 4 °C or 5 °C would be far worse. It would be tragic if 30 percent of Earth's existing species went extinct, but it would be far more tragic if 50 percent of Earth's existing species went extinct. It would be an unprecedented global disaster if climate change caused 300 million casualties, but it would be an even greater disaster if it led to 500 million casualties.

Asking whether we can solve the problem misconstrues our situation. The better question is this: *what can we do to make things better*? While loss is inevitable, the scope and severity of that loss can be mitigated. Acting as soon as possible is the path to making things better; delaying action will make things worse.

Making progress on population growth

In this book, I have focused on ways we might take action with respect to population growth (though obviously a comprehensive response will require focusing on more than just that issue). Admittedly, population remains off limits as a discussion topic in many political circles, and serious political action to curtail it is unlikely in the near future. Some population activists, such as Dave Foreman (2014) and Karen Shragg (2015), offer advice about how individuals might change those circumstances by promoting awareness or getting involved with organizations that are trying to reduce population growth. Sometimes activists suggest additional political actions, such as calling a local representative or advocating for a cap on population growth in your hometown. Measures like these can certainly play a role in trying to build a consensus among the public that population needs to be taken seriously, but an adequate response to population growth will require coordinated international action that is unlikely to materialize in the immediate future.

Nevertheless, there are some reasons to be optimistic. Over the last decade, academics have started to take population seriously again. (The references throughout this text serve as evidence of that.) Moreover, recent articles in several news outlets and magazines have explicitly engaged with how climate change may the affect the morality of procreation (Astor 2018; Irfan 2019; Kirkey 2019; Paddison 2019; Papazoglou 2019). These articles demonstrate that concerns about population and procreation are not limited to specialized academics and that many people are starting to think more reflectively about their procreative choices. Additionally, in the world's most developed nations, there is a growing trend of couples voluntarily choosing to go childless (Blackstone 2019). As this decision becomes more common, it will become more socially acceptable, and people will be under less pressure to have children that they do not really want.

As time passes, the population problem will also become more pronounced, and this will force people to confront it more directly. We are already well above the number of people that the Earth can support in the long term. Eventually, we will have to take action to reduce our numbers or else face devastating consequences. But that choice has not yet imposed itself on us with enough urgency to compel political action. Nonetheless, progress can still be made in the short term by thinking through the challenges that lay ahead and confronting the reality of our circumstances. After all, even some of those who know the facts about population retreat from discussing the topic. When Alan Weisman interviewed *haredi* environmental educator Rachel Ladani about what will happen in 2050 when Israel has twice its current population and the global population nears ten billion, she replied, "I don't have to think about it. God made the problem, and He will solve it" (Weisman 2013, p. 12). While a miracle or two might well help us respond to these environmental problems, expecting divine intervention would be foolhardy. We will have to cope with these problems on our own.

Not everyone tries to escape our environmental problems by retreating to religion. Others, as discussed in Chapter 4, put their faith in technological progress. Others simply deny the problems exist. In finishing this book, my hope is that you will have a different reaction. I hope that you will appreciate the gravity of our situation without losing heart or disconnecting from the issue. Neither languishing in despair nor hiding from the problem will accomplish anything. There are more of us on the planet every day, and the sooner we confront the moral significance of this fact, the better off we will be.

Notes

1 For two of his recent and representative books on this subject, see McPherson (2019a, 2019b).
2 McPherson also isn't the only environmental writer to endorse an apocalyptic perspective. For a brief overview of McPherson and some other environmental doomsayers, see Wallace-Wells (2019, pp. 204–216).
3 For an overview of this phenomenon (commonly called "implicit bias"), see Brownstein (2019).
4 Eliminating racism entirely under current conditions would probably require widespread neural modification in how our brains make subconscious associations. Such technology is not currently available, and its potential use would raise a separate set of ethical concerns.
5 For those who want a different and more comprehensive response to the phenomenon of environmental despair, see McKinnon (2014).

References

Astor, Maggie. 2018. "No Children Because of Climate Change? Some People Are Considering It." *New York Times*. www.nytimes.com/2018/02/05/climate/climate-change-children.html. Accessed November 10, 2019.

Blackstone, Amy. 2019. *Childfree by Choice: The Movement Redefining Family and Creating a New Age of Independence*. New York: Dutton.

Brownstein, Michael. 2019. "Implicit Bias." *Stanford Encyclopedia of Philosophy*. http://plato.stanford.edu/entries/implicit-bias/. Accessed November 10, 2019.

Foreman, Dave. 2014. *Man Swarm: How Overpopulation Is Killing the Wild World*, 2nd ed., edited by Laura Carroll. LiveTrue Books.

Irfan, Umair. 2019. "We Need to Talk about the Ethics of Having Children in a Warming World." *Vox*. www.vox.com/2019/3/11/18256166/climate-change-having-kids. Accessed November 10, 2019.

Kirkey, Sharon. 2019. "Is It Immoral to Have Babies in the Era of Climate Change?" *National Post*. https://nationalpost.com/life/is-it-immoral-to-have-babies-in-the-era-of-climate-change. Accessed November 10, 2019.

McKinnon, Catriona. 2014. "Climate Change: Against Despair." *Ethics and the Environment* 19, no. 1: 31–48.

McPherson, Guy. 2016. "Climate-Change Summary and Update." https://guymcpherson.com/climate-chaos/climate-change-summary-and-update/. Accessed on November 10, 2019.

McPherson, Guy. 2019a. *Going Dark*, 2nd ed. Pleasantville, NY: Wood Thrush Productions.

McPherson, Guy. 2019b. *Only Love Remains: Dancing at the Edge of Extinction*. Pleasantville, NY: Wood Thrush Productions.

Paddison, Laura. 2019. "9 People on the Ethics of Having Kids in an Era of Climate Crisis." *Huffpost*. www.huffpost.com/entry/climate-change-having-kids-children_n_5d493eaee4b0244052e09033. Accessed November 10, 2019.

Papazoglou, Alexis. 2019. "Is It Cruel to Have Kids in the Era of Climate Change?" *The New Republic*. https://newrepublic.com/article/153149/cruel-kids-era-climate-change. Accessed November 10, 2019.

Shragg, Karen. 2015. *Move Upstream: A Call to Solve Overpopulation.* Minneapolis–St. Paul, MN: Freethought House.
Wallace-Wells, David. 2019. *The Uninhabitable Earth: Life After Warming.* New York: Tim Duggan Books.
Weisman, Alan. 2013. *Countdown: Our Last, Best Hope for a Future on Earth?* New York: Little, Brown, and Company.

Appendix
The non-identity problem

The non-identity problem (NIP) refers to the concern that certain future people cannot be harmed because their identities can be altered by the actions we perform. In Chapter 3, I dismissed NIP on two grounds. First, I highlighted the fact that many of the environmental problems under consideration will affect large numbers of people who already exist or will exist regardless of what we do in the near future. Since these people's identities will not be altered by our actions, they will be made worse off by these environmental harms than they otherwise would have been. As a result, NIP doesn't provide grounds for thinking these people will not be harmed. Second, I suggested that the counterfactual-comparative account of harm that underlies the non-identity problem should be rejected. Intergenerational moral concerns reveal that this account of harm is too narrow in scope to serve as the sole basis for our duties of non-harm. I kept my remarks on NIP brief in that chapter to avoid a lengthy, theoretical digression from the chapter's central argument. This appendix provides a more thorough examination of NIP to satisfy those who view it as a dire threat to intergenerational ethics.

Here, I examine NIP as presented by David Boonin (2014) because he has offered the most comprehensive defense of NIP to date. His illustration of the problem focuses on a particular procreative decision (Boonin 2014, p. 2).[1] Wilma is considering whether or not to have a baby. If she conceives a child in the immediate future, this child will suffer from a severe and irreversible disability. However, the condition would not be so severe as to render the child's life not worth living. Fortunately, this outcome is preventable: if Wilma takes a small pill once a day for two months before conceiving a child, then her child will be born without this disability. The costs of the medication will be covered by Wilma's health insurance, and the pill has no side effects. Yet, despite these facts, Wilma chooses to conceive a child immediately, and her child Pebbles is born with incurable blindness.

I suspect that most who read this case will regard Wilma's decision as morally wrong: she conceived a blind child when she could have conceived a child with normal vision at little cost to herself. That seems wrong, but if Wilma had taken the pill, then the child she conceived would not have been Pebbles – a different child would have been born instead. After two months, the sperm and egg pair that united in conception would be completely different, and since the child would have had a different genetic constitution than Pebbles, that child would not be Pebbles. Since it is impossible for Pebbles to exist and not be blind, she is not made worse off than she otherwise would have been through Wilma's act of conceiving her. As a result, we must wonder whether Wilma actually did something wrong in conceiving her.

Boonin (2014) uses Wilma's procreative decision as a backdrop for articulating what he calls the Non-Identity Argument (p. 27):

P1: Wilma's act of conceiving now rather than taking a pill once a day for two months before conceiving does not make Pebbles worse off than she would otherwise have been.
P2: If A's act harms B, then A's act makes B worse off than B would otherwise have been.
P3: Wilma's act of conceiving now rather than taking a pill once a day for two months before conceiving does not harm anyone other than Pebbles.
P4: If an act does not harm anyone, then the act does not wrong anyone.
P5: If an act does not wrong anyone, then the act is not morally wrong.
C: Wilma's act of conceiving Pebbles is not morally wrong.

The first five claims represent the argument's premises, and the final claim represents the argument's conclusion. The way that the argument is structured, accepting all the premises as true will entail that we must accept the conclusion that Wilma did not act wrongly in conceiving Pebbles. Thus, the main question is whether or not there is a premise in the argument that we should reject.

As my earlier remarks suggest, I favor rejecting the second premise of the argument, but in what follows, I also argue that the fourth and fifth premises are false. Specifically, I argue that it is possible for a person to be harmed even in some cases where the person is not made worse off than they otherwise would have been, that it is possible to wrong a person without causing them harm, and that actions can be morally wrong even when they do not wrong any particular person.

The counterfactual comparative account of harm

While I have already addressed P2 of the Non-Identity Argument, many philosophers are drawn to the account of harm that it represents. So I will start by revisiting that claim and offering a more thorough appraisal of it. Again, here is the relevant claim: *If A's act harms B, then A's act makes B worse off than B would otherwise have been.* This statement reflects what is sometimes called the counterfactual comparative account of harm (CCH). The standard defense of CCH is that it accords with the commonsense understanding of what it means to harm someone. On this basis, Boonin (2014) presents CCH as the default position "unless a better alternative comes along" (p. 52).

Boonin bolsters his defense of P2 by providing some anecdotes about how people could explain how they were harmed or justify claims that they in fact did not harm someone else. Imagine if I were to vandalize your car. How would you explain why you were harmed? Boonin (2014) suggests that your likely reply would be "pointing to the various ways in which my act has made you worse off than you would have been had I not vandalized your car" (p. 52). If I asked you how much money I needed to pay you to compensate for the loss, you would presumably suggest the amount that comes closest to nullifying the extent that my vandalism has made you worse off – perhaps the total cost of both repairing the vehicle and renting a replacement vehicle while those repairs are done. Furthermore, if you were asked to explain why you did not harm me when you scratched your nose, then "you are likely to appeal to the claim that your act did not in any way make me worse off than I would otherwise have been" (p. 52).

These anecdotes sound plausible upon an initial reading, but this general strategy of appealing to moral common sense has two serious shortcomings. First, it is questionable whether common sense is an appropriate starting point for defending abstract theoretical claims about morality. History is littered with cases where a population's commonsense beliefs about morality, such as beliefs that race-based slavery was justified or that women should not have the same civil and political liberties as men, are now regarded as deeply mistaken. If commonsense morality is a suitable starting point for assessing moral claims, then they should at least be supported by some plausible line of reasoning as well.

The second shortcoming of an appeal to commonsense morality is that one has to demonstrate that the claim in question is in fact part of commonsense morality. The most basic way to do this would be to conduct a poll to see if a particular moral claim is in fact a part of most people's moral beliefs. Boonin has not made an attempt to conduct this empirical test, and to my knowledge, no one else has either. As a result, it is just speculation that CCH is a part of commonsense morality. Why should this be assumed? While harm is probably a component of almost everyone's moral reasoning, I suspect that most ordinary people do not spend much time reflecting on the necessary and sufficient conditions for harming someone. I imagine that many people's understanding of harm is an intertwined morass of inconsistent intuitions that does not converge on any specific account of harm. Until we encounter evidence that people really do converge on CCH in the way Boonin suggests, we are not justified in assuming that CCH is a part of commonsense morality.

My reasoning thus far has tried to undercut some of the support for CCH, but I have not yet provided a compelling reason to reject it. So let's look at some deeper problems with CCH. One concern is that this account of harm cannot handle cases of preemption. Here is a preemption case that comes from Hanser (2008):

The Two Hit Men: Mr. Bad orders Hit Man 1 to shoot and kill you. Hit Man 1 doesn't always follow his orders, so Mr. Bad orders Hit Man 2 to shoot and kill you if Hit Man 1 fails to shoot and kill you. Hit Man 2 always follows his orders. As it happens, though, Hit Man 1 shoots and kills you and Hit Man 2's orders prove to be unnecessary.

(p. 436)[2]

By shooting and killing you, Hit Man 1 does not make you worse off than you otherwise would have been. If he had not followed his orders, then Hit Man 2 would have shot and killed you, leading to the same result. So, according to CCH, Hit Man 1 did not harm you. Such a result is mind-boggling: it means that an act of murder no longer constitutes a harm so long as another person was about to commit the same murder! Boonin (2014) recognizes this problematic implication but nonetheless claims that a proponent of CCH should bite the bullet and accept this outcome (p. 58). I favor a different strategy: reject CCH. Preemption cases are scenarios where a person can be harmed even though they are not made worse off than they otherwise would have been.

Scenarios involving omissions are also sometimes presented as problem cases for CCH. Imagine that Batman purchases some golf clubs, intending to give them to Robin, but he ultimately decides to keep the clubs for himself. Since receiving these clubs would make Robin happy, it appears that Batman's keeping them makes Robin worse off than he otherwise would have been (Bradley 2012, p. 397). Such an implication would be worrying because it suggests that you can harm another person simply by deciding not to provide a benefit to them. But failures to benefit someone are not typically regarded as identical to harming that person. Suppose that you intend to buy your friend a gift at one time but later decide not to. If you share your decision-making process with your friend later, do you think they will insist that you harmed them by deciding not to purchase that gift? I suspect such a thought wouldn't even cross their mind.

The literature contains a surplus of discussions about preemption, omission, and CCH (e.g., Bradley 2012; de Villers-Botha 2018; Feit 2015; Hanna 2016; Hanser 2008; Johansson and Risberg 2019; Klocksiem 2012). I am pessimistic about attempts to revise CCH to avoid these difficulties,[3] and since I do not consider CCH to be an entrenched part of commonsense morality, I am not motivated to stick with it in the way that philosophers like Boonin do.

As one would expect, Boonin is aware of these concerns with CCH, and his main response is to point out that its competitors have even bigger problems. However, this strategy is misguided: even if we established that CCH was better than all rival accounts of harm, that would not establish that CCH was *true*. We might instead conclude that harm is a multifaceted concept that cannot be reduced to a single unified theory.[4] Regardless, we do not need to establish a full account of harm to reject the second premise of the Non-Identity Argument. We only need to establish that CCH should be rejected, and I think we have more than enough reason to abandon this account: a person can be harmed without being made worse off than they otherwise would have been.

Wrongful acts that do not cause harm

In the prior section, I argued that the Non-Identity Argument should be rejected because CCH is false. But let's suppose that solutions for the problems associated with this account of harm are eventually found. Even under those circumstances, the Non-Identity Argument would not succeed because two of its other premises are false. One of these false claims is the fourth premise (P4): *if an act does not harm anyone, then the act does not wrong anyone.* The most straightforward way to challenge this premise is to present counterexamples, and there are many good candidates.

Some counterexamples to this premise are cases where a person's rights are violated without the person being harmed. Acts of voyeurism violate a person's right to privacy, and on these grounds, they are wrong even when they are never discovered and cause no one any harm. Sexual assaulting an unconscious person violates a person's right to bodily integrity, and thus, this action is wrong even if no one else learns about the act and it causes no physical harm to the victim.

Another type of counterexample is non-harmful promise-breaking. Even if you can break a promise without causing harm and without this fact becoming known, you are not obviously permitted to break your promise. Imagine making a deathbed promise to one of your parents that you will try to reconcile a fractured relationship with one of your siblings. This promise was made in private, so no one else knows about it. No one will be harmed by your breaking your promise, and your attempts to resolve the problems between you and your sibling are unlikely to succeed anyway. Yet these facts alone do not make it clear that you would be justified in ignoring the promise that you made. After all, the practice of promise-making is grounded in the idea that these agreements should be upheld except in dire circumstances.

Finally, there are counterexamples that are grounded in the mere *risk* of harm. It is morally wrong to drive drunk even when no harm occurs because the risk of causing harm to someone else is unacceptably high. The wrongness of this action does not hinge on whether or not a particular instance of it causes harm. Whether someone is injured by a drunk driver typically involves a good deal of luck, such as when the driver leaves the parking lot or where on the sidewalk a pedestrian happens to be standing when the driver approaches.[5] The wrongness of one's conduct cannot hinge on luck in this manner. If sheer luck could make such a dramatic difference in our level of moral responsibility, then we would not have much control over our moral conduct, and judgments about our moral character would be arbitrary: being a good person would be primarily a result of one's good fortune rather than a result of cultivating virtuous dispositions or engaging in careful moral deliberation.[6]

As one might expect, Boonin is aware of these potential objections to P4. His responds by noting that these concerns do not explain how Wilma's specific act is wrong and thinks it is possible to modify P4 to avoid these shortcomings. Thus, as he puts it, "the question is not whether we can show that P4 is false. The question is whether we can show that it is false enough" (Boonin 2014, p. 109).[7] While this is a fair point, I believe we have the argumentative resources to show this premise "false enough" to undermine the Non-Identity Argument.

I earlier mentioned counterexamples to P4 that are based on rights violations, and this is the best route for arguing that Wilma's specific action is wrong. The most plausible candidate for a right that Wilma violates is a child's right "not to be born with important opportunities foreclosed" (Jecker 2012, p. 34).[8] This right explains why intentionally having a child who will be blind (or suffer from some other significant disability) is usually wrong. It also explains why it is often wrong to have children when they do not have a favorable chance at a good life, such as when the parents lack the means to provide the child adequate care. Even if we assume that Pebbles has not been harmed, she has been wronged because her right to non-foreclosed opportunities has been violated.[9]

One crucial aspect of Wilma's situation is that she can, at a rather low cost to herself, conceive a child who will not endure this rights violation. If it were not possible for Wilma to have a biological child who would be sighted, then we might reason that her action is permissible despite the rights violation that would occur. Rights violations are *prima facie* morally wrong, but they are not *always* wrong. An inability to conceive a sighted child might serve as a sufficient justification for violating this right – at least if Wilma is able to give her blind child a favorable chance at a good life.[10] Nonetheless, there is a strong presumption against violating a child's right to non-foreclosed opportunities, and a trivial inconvenience – in Wilma's case, taking a pill daily for two months – is not sufficient to justify violating this right.

Intriguingly, Boonin (2014) doubts that what happens to Pebbles can be considered a rights violation (pp. 111–113). He suggests that the only strategy one could use to defend is view is to claim that Pebbles has a right not to exist in her present condition, a tactic he attributes to Doran Smolkin (1999). This view is broadly consistent with the right of non-foreclosed opportunities since Pebbles' present condition does entail having various opportunities eliminated from her future.

Boonin's (2014) main objection to this strategy is that it violates what he calls the "Independence Requirement" (pp. 20–21). This requirement is one of the conditions that he thinks a satisfactory response to the Non-Identity Argument must meet. The Independence Requirement states that any reason for rejecting a premise in the Non-Identity Argument must be independent of the fact that rejecting that premise would enable us to avoid the argument's conclusion. In other words, an objection to one of the premises must be motivated by a consideration other than reluctance to accept that Wilma's act of conceiving Pebbles was permissible.

Boonin aims this response primarily at Smolkin's (1999) presentation of a direct rights-based argument, and admittedly, Smolkin offers little to motivate his rights-based account other than its ability to avoid NIP. However, the right not to have one's opportunities foreclosed does not suffer from the same shortcoming: this right explains why it is wrong to have children in circumstances where they will not have a favorable chance at a good life, so it has some plausible support that is independent of its ability to avoid NIP. Consequently, his response fails, and we have a good reason to reject P4. Wilma's conceiving Pebbles was morally wrong even if we assume that her action did not cause Pebbles harm.

Vicious parenting

I regard the preceding flaws in the Non-Identity Argument to be sufficient grounds for rejecting it, but for readers who are still uncertain, there are also good reasons to reject P5. According to this premise, *an act is not morally wrong if it does not wrong anyone*. So, for this short section, let's assume that Wilma's action neither harms nor wrongs Pebbles or anyone else. I contend that there are still moral grounds for opposing her action. To make this case, we will draw on the moral tradition of virtue ethics.

Virtue ethics focuses on a person's moral character. This moral tradition does not fixate on the rightness or wrongness of particular actions. Instead, it considers our dispositions to think, feel, and behave in certain ways. Now imagine what ideally virtuous parents would be like. These parents would deeply love and cherish their child, and they would strive to promote their child's welfare while also promoting and respecting their child's autonomy. Now imagine that one of these parents is in Wilma's position. Would a virtuous parent choose to conceive her child immediately instead of taking the pill for two months and then conceiving? The answer is surely no, and in this observation, we find the basis for an objection to P5.

Even if we assume that Pebbles' birth does not harm or wrong anyone, conceiving Pebbles is still wrong because it instantiates a vice. Wilma is not sufficiently concerned with providing her future child the best possible chance at a good life – something a virtuous parent will care about regardless of who their child happens to be. Wilma's indifference to her future child's welfare and life prospects represents a failing of her moral character. While parents are not typically expected to do everything possible to maximize their child's chances of living a good life,[11] they are typically expected to prioritize their child's prospects for living well and act accordingly when they can do so at relatively low costs to themselves. Taking a pill daily for two months is not a heavy burden, so Wilma's choice to conceive immediately is not compatible with being a virtuous parent.

Boonin (2014) briefly acknowledges this virtue-based objection to P5 (pp. 184–188), but he characterizes the relevant vice as insensitivity to suffering.[12] The relevant character flaw, however, is not necessarily insensitivity to suffering. We can understand this vice as a kind of parental negligence – a failure to care sufficiently about the impersonal welfare of one's children. David DeGrazia (2012) adopts a similar stance when he evaluates a case in his own work that is very similar to Wilma's:

> ... it is very clear that the parents did not make this choice *in order to benefit this very child*. Indeed, their conduct expressed a highly cavalier attitude about their procreative options and their likely consequences. In this way, the parents expressed a profound lack of regard for *their offspring—whoever it would be*.... Although the couples' disregard was not intentionally directed at the child they had, it was, in a sense, negligently directed at whatever child they might have.
>
> (pp. 180–181, original emphasis)

Wilma should care about the circumstances of her child's birth – whoever her child happens to be and should avoid actions that lower the impersonal welfare of her child for trivial reasons. But she does not deliberate this way or act accordingly.[13] Her behavior is an instance of vicious parenting, and therefore, it is wrong. P5 should be rejected.

Rejecting the non-identity argument

The Non-Identity Argument threatens to raise some serious challenges to how we approach moral problems in intergenerational and procreative ethics, but the argument is unsuccessful. The argument's second, fourth, and fifth premises are false. Nevertheless, since NIP has occupied such a prominent place in philosophical discussions of population ethics in the last few decades, I will close with two general notes for readers who want to think more about this subject.

First, the scope of NIP is surprisingly narrow, a fact highlighted most forcefully by Rivka Weinberg (2016). NIP applies only to narrow, person-affecting moral theories. These theories evaluate whether actions are right or wrong based on how they affect particular identifiable individuals. Some moral theories clearly do not fit this description. Consequentialist theories, for instance, evaluate whether actions are right or wrong based on their overall consequences. Since these theories are concerned with aggregate goodness (or badness), they are not narrow, person-affecting theories. Virtue ethical theories are also not narrow, person-affecting theories. This moral tradition focuses on developing virtuous character and acting in accordance with this character. So what makes an action right or wrong does not depend on the particular identity of the individual affected: what matters is whether or not the action properly followed from virtuous character.

NIP is usually presented as applying to deontological moral theories, but even many deontological theories are not narrowly person-affecting in the way that NIP requires. Deontological moral theories typically determine the permissibility of an action based on whether or not that action is consistent with particular moral principles – usually principles that are designed to respect people's moral agency and treat them as ends-in-themselves. As Weinberg (2016) points out, NIP only arises "if permissibility of acts is determined by the act's *effects* or consequences on a particular person" (p. 105, original emphasis). NIP is not applicable to most deontological theories because they are often *not* concerned with a specific action's effects on a particular person. Instead, they are based on adherence to moral principles regardless of a specific action's consequences. For these reasons, most deontological theories will not be consistent with P4 of the Non-Identity Argument. NIP will only be applicable to the subset of deontological moral theories that adopt the specific notions of harm and wrongness that it presupposes.

My second note about NIP is that my analysis here only scratches the surface of potential responses. It has been addressed hundreds of times within the last three decades, and there are dozens of proposed solutions.[14] Philosophers frequently disagree about the best way to respond to NIP, but there is a broad consensus that it can be resolved. Despite how often NIP is invoked in philosophical discussion about intergenerational ethics, it is not a significant obstacle to reasoning about what we owe to future people.[15]

Notes

1 Boonin's case has some significant structural parallels to Parfit's (1982) *Handicapped Child* (p. 118).
2 For some other examples of preemption cases that have been raised as problems for CCH, see Thomson (2011, pp. 446–447), Woollard (2012, p. 484), and Bradley (2012, p. 397).
3 These difficulties are also not the only potential problems for CCH. There are concerns about whether it can adequately account for the harm of death (Hanser 2008, 2011; Purves 2016) and whether it is compatible with the prudential and moral relevance of harm and benefit (Carlson 2019). NIP is also frequently mentioned as a counterexample to CCH, but it would obviously beg the question to present it as such in this context. We would have to assume that the Non-Identity Argument is unsound to press that claim, and the argument's soundness is precisely what we are trying to assess.
4 Ben Bradley (2012) appears to reach a similar conclusion, but he also expresses pessimism about the usefulness of the concept of harm under such circumstances: "My diagnosis is that the notion of harm is a Frankensteinian jumble. Thus it is unsuitable for use in serious moral theorizing" (p. 391). I agree with his assessment of the concept of harm, but I do not share his skepticism about its usefulness in moral reasoning. As I mentioned at the outset of Chapter 3, a duty to avoid unnecessary harm is one of the most basic and uncontroversial moral principles.

5 Nagel (1979) discusses drunk driving in the context of moral luck. I borrow the connection to risk of harm from Kumar (2003, p. 103).
6 For an overview of the ways in which moral luck remains a challenge for moral theorizing, see Nelkin (2019).
7 Perhaps surprisingly, Boonin does not ever present a version of the fourth premise that is immune to all these counterexamples but still upholds the conclusion of the Non-Identity Argument. He leaves that task to the reader.
8 Velleman (2008) proposes something similar, suggesting that children have "a right to be born into good enough circumstances" (p. 275). Presumably, "good enough circumstances" would include the condition that the child has not had important opportunities foreclosed.
9 Here, I adopt the view that rights apply universally to all people and that they specify certain moral thresholds below which no one should sink. This outlook on rights is similar to the one endorsed by Simon Caney (2010), who identifies human rights as "minimum moral thresholds to which all individuals are entitled, simply by virtue of their humanity, and which override all other moral values" (p. 165). See also Shue (1996) and Bell (2011, pp. 104–110). Thus, Pebbles has had her rights violated despite the fact that she could not have existed without that rights violation occurring: Wilma's action has caused her to sink below a moral threshold that no one – regardless of their identity – should sink beneath.
10 If we regard the right to non-foreclosed opportunities as particularly stringent, then we might also reject this line of reasoning and think Wilma's action is wrong even if it's the only way she can have a child of her own. On that line of reasoning, she would be obligated to pursue adoption or use a surrogate mother if she wanted to raise a child.
11 One exception is Julian Savulescu (2001). He defends the principle of Procreative Beneficence: "couples (or single reproducers) should select the child of the possible children they could have, who is expected to have the best life, or at least as good a life as the others, based on the relevant, available information" (p. 415). See also Savulescu (2007), and Savulescu and Kahane (2009). Those who endorse this position will obviously object to Wilma's behavior, but we do not need to endorse such a strong position to reach that conclusion.
12 Boonin focuses primarily on how this vice is discussed by Urbanek (2010).
13 This line of reasoning sounds rather consequentialist in nature, but one does not need to endorse strict consequentialism to accept it. Any plausible account of morality will consider the impact that our actions have on the welfare of others, even if it is true that impersonal welfare is not the *only* thing that matters. The general point is that parents should strive not to lower the impersonal welfare of their children without good reason.
14 For some responses that I have not examined here, see Woodward (1986), 't Hooft (1999, pp. 50–51), Roberts (2007), Davidson (2008, p. 482), Harman (2009), Nolt (2011, pp. 71–72), Das (2014), Weinberg (2016, ch. 3), Kumar (2018), and Meyer (2018).
15 This appendix combines some responses to NIP that I have articulated in the past. See Hedberg (2013, pp. 11–25, 2017, pp. 35–46).

References

Bell, Derek. 2011. "Does Anthropogenic Climate Change Violate Human Rights?" *Critical Review of International Social and Political Philosophy* 14, no. 2: 99–124.

Boonin, David. 2014. *The Non-Identity Problem and the Ethics of Future People*. Oxford: Oxford University Press.

Bradley, Ben. 2012. "Doing Away with Harm." *Philosophy and Phenomenological Research* 85, no. 2: 390–412.

Caney, Simon. 2010. "Climate Change, Human Rights, and Moral Thresholds." In *Climate Ethics: Essential Readings*, edited by Stephen Gardiner, et al., pp. 163–177. New York: Oxford University Press.

Carlson, Erik. 2019. "More Problems for the Counterfactual Comparative Account of Harm and Benefit." *Ethical Theory and Moral Practice* 22, no. 4: 795–807.

Das, Ramon. 2014. "Has Industrialization Benefited No One? Climate Change and the Non-Identity Problem." *Ethical Theory and Moral Practice* 17, no. 4: 747–759.

Davidson, Marc. 2008. "Wrongful Harm to Future Generations: The Case of Climate Change." *Environmental Values* 17: 471–488.

DeGrazia, David. 2012. *Creation Ethics: Reproduction, Genetics, and Quality of Life*. Oxford: Oxford University Press.

de Villers-Botha, Tanya. 2018. "Harm: The Counterfactual Comparative Account, the Omission and Pre-emption Problems, and Well-being." *South African Journal of Philosophy* 37, no. 1: 1–17.

Feit, Neil. 2015. "Plural Harm." *Philosophy and Phenomenological Research* 90, no. 2: 361–388.

Hanna, Nathan. 2016. "Harm: Omission, Preemption, and Freedom." *Philosophy and Phenomenological Research* 93, no. 2: 251–273.

Hanser, Matthew. 2008. "The Metaphysics of Harm." *Philosophy and Phenomenological Research* 77, no. 2: 421–450.

Hanser, Matthew. 2011. "Still More on the Metaphysics of Harm." *Philosophy and Phenomenological Research* 82, no. 2: 459–469.

Harman, Elizabeth. 2009. "Harming as Causing Harm." In *Harming Future Persons*, edited by Melinda Roberts and David Wasserman, pp. 137–154. Dordrecht, Netherlands: Springer.

Hedberg, Trevor. 2013. "Wouldn't Future People Like to Know? A Compensation-Based Approach to Global Climate Change." M.A. thesis, Department of Philosophy, University of Tennessee.

Hedberg, Trevor. 2017. "Population, Consumption, and Procreation: Ethical Implications for Humanity's Future." Ph.D. dissertation, Department of Philosophy, University of Tennessee.

Jecker, Nancy. 2012. "The Right Not to Be Born: Reinterpreting the Nonidentity Problem." *American Journal of Bioethics* 12, no. 8: 34–35.

Johansson, Jens, and Olle Risberg. 2019. "The Preemption Problem." *Philosophical Studies* 176, no. 2: 351–365.

Klocksiem, Justin. 2012. "A Defense of the Counterfactual Comparative Account of Harm." *American Philosophical Quarterly* 49, no. 4: 285–300.

Kumar, Rahul. 2003. "Who Can Be Wronged?" *Philosophy & Public Affairs* 31, no. 2: 99–118.

Kumar, Rahul. 2018. "Risking Future Generations." *Ethical Theory and Moral Practice* 21, no. 2: 245–257.

Meyer, Kirsten. 2018. "The Claims of Future Persons." *Erkenntnis* 83, no. 1: 43–59.

Nagel, Thomas. 1979. "Moral Luck." *Mortal Questions*, pp. 24–38. Cambridge: Cambridge University Press.

Nelkin, Dana. 2019. "Moral Luck." *Stanford Encyclopedia of Philosophy*. http://plato.stanford.edu/entries/moral-luck/. Accessed October 27, 2019.

Nolt, John. 2011. "Greenhouse Gas Emissions and the Domination of Posterity." In *The Ethics of Global Climate Change*, edited by Denis Arnold, 60–76. Cambridge: Cambridge University Press.

Parfit, Derek. 1982. "Future Generations, Further Problems." *Philosophy & Public Affairs* 11, no. 2: 113–172.

Purves, Duncan. 2016. "Accounting for the Harm of Death." *Pacific Philosophical Quarterly* 97, no 1: 89–112.

Roberts, Melinda. 2007. "The Non-identity Fallacy: Harm, Probability, and Another Look at Parfit's Depletion Example." *Utilitas* 19, no. 3: 267–311.

Savulescu, Julian. 2001. "Procreative Beneficence: Why We Should Select the Best Children." *Bioethics* 15, no. 5/6: 414–426.

Savulescu, Julian. 2007. "In Defence of Procreative Beneficence." *Journal of Medical Ethics* 33, no. 5: 284–288.

Savulescu, Julian, and Guy Kahane. 2009. "The Moral Obligation to Create Children with the Best Chance of the Best Life." *Bioethics* 23, no. 5: 274–290.

Shue, Henry. 1996. *Basic Rights: Subsistence, Affluence, and U.S. Foreign Policy*, 2nd ed. Princeton: Princeton University Press.

Smolkin, Doran. 1999. "Towards a Rights-Based Solution to the Non-Identity Problem." *Journal of Social Philosophy* 30, no. 1: 194–208.

Thomson, Judith. 2011. "More on the Metaphysics of Harm." *Philosophy and Phenomenological Research* 82, no. 2: 436–458.

't Hooft, Visser. 1999. *Justice to Future Generations and the Environment*. Dordrecht: Luwer Academic Publishers.

Urbanek, Valentina Maria. 2010. "The Non-Identity Problem." Ph.D. dissertation, Department of Linguistics and Philosophy, MIT.

Velleman, J. David. 2008. "Persons in Prospect, Part III: Love and Nonexistence." *Philosophy and Public Affairs* 36, no. 3: 266–288.

Weinberg, Rivka. 2016. *The Risk of a Lifetime: How, When, and Why Procreation May Be Permissible*. New York: Oxford University Press.

Woodward, James. 1986. "The Nonidentity Problem." *Ethics* 96, no. 4: 804–831.

Woollard, Fiona. 2012. "Have We Solved the Non-Identity Problem?" *Ethical Theory and Moral Practice* 15, no. 5: 677–690.

Index

abortion 5, 56, 67–70; criminalization of 137; forced 63, 74, 131, 137
adoption 97, 121–122, 127n13, 133
anthropocentrism 151, 156
animal agriculture 153–155; contribution to climate change 157n3
antinatalism 109–110, 135–136; axiological asymmetry argument for 110–111; conditional 110, 121–125; consent-based arguments for 116–119; misanthropic argument for 115–116; quality-of-life argument for 111–115; risk aversion argument for 119–121; unconditional 110–121

Benatar, David 110–116
biocentrism 155
biodiversity: definition of 18; loss 18–19, 22–23, 58, 156, 163; value of 19–22
Bognar, Greg 72
Boonin, David 167–170, 172–174
Broome, John 85
Burning Building 43–44

carbon footprint 84–85, 90–91, 93; of procreation 54, 87–88
carrying capacity 15
Campbell, Martha 2–3
climate change 16–18, 58, 163; and biodiversity loss 18, 156; casualties caused by 16–17, 56; and the Doomsday Clock 23; duration of 17–18; and immigration 146–147; mitigation of 52–53; and sea level rise 18; *see also* greenhouse gas emissions
coercion 63–64, 69–71, 73–75
Conly, Sarah 74–75, 132–134
consciousness 68; in animals 151–153, 157

consequentialism 7, 12n8, 174, 176n13; of rights 77–78
considered judgments 8–9; *see also* reflective equilibrium
contraception *see* family planning
Crist, Eileen 146

demandingness 94, 96–97, 123–124
demographic transition theory 3
deontology 175
Depletion 39–40
desire fulfillment 112–114

Equity of Non-Harm 35–39
Erlich, Paul 14–15
extinction: of humans 161–162; of nonhuman species *see* biodiversity, loss

factory farms 153–154
family planning 5–6, 55–56, 64–67, 70, 72–73, 76–78, 138; and abortion 69
fertility rate 4–5; lowering of 14, 51, 55, 63, 66–67, 69–70, 100; and one-child policies 74
food shortage 15
Friedrich, Daniel 121–122
future people: definition of 35; compared to merely possible people 35–36, 118–119, 138n5

gender justice 65–67
geoengineering 17, 58
greenhouse gas emissions: global 52–54; caused by procreation 54; and fertility rates 55; and individual obligations 84–93; *see also* climate change

habitat destruction 19, 155
harm 34–35; to animals 153–155, 157; and antinatalism 110–111; caused by lifetime

harm *continued*
  greenhouse gas emissions 85–86; and consent 116–119, 125; counterfactual comparative account of 39–41, 43–45, 169–171; to future people 35–39; massive systematic 87–88; of procreation 133–134; risk of 117–121, 125, 171
Häyry, Matti 119–120, 127n10
hedonism 112–114

immigration 143–144; and climate change 146–147; and domestic population growth 145–146; and economic growth 147–148
incentivization 69–73
integrity 88, 102n11; value of 88–90; and duty to reduce greenhouse gas emissions 90–92, 102n13; and duty to reduce procreation 93–94
IPAT equation 50

Klein, Naomi 52
Kukla, Rebecca 77, 135–136

Licon, Jimmy 118–119
life plans 114; role of children in 93–94, 96, 121–122, 124–125, 127n16

Maier, Don 20–22
Marquis, Don 68–69
Malthus, Thomas 14–15
maximin rule 119–120
McPherson, Guy 161–162
moral cosmopolitanism 144–145
moral tragedy 75–76; and immigration 146–147

Nolt, John 85
non-identity problem 39–41, 167–168, 174–175
Non-Identity Argument 168; Independence Requirement of 172–173

offsetting 99–100
one-child policy 74–75
optimism bias 112
Overall, Christine 94–96
overconsumption 2, 50

Parfit, Derek 39–40
personhood 67–69
Polluter Pays principle 80n9
population: activism 163–164; and animal suffering 154–155; control 5, 52, 137–138; growth 1, 14, 51, 59, 137; past discussion of 2–7; policy and inequality 77; relation to consumption 6–7, 50, 53–55; relation to economic growth 76, 147; relation to health care 76; and technological progress 55–59
Population Bomb, The 14
Population Reduction Argument 33–34, 49, 59
poverty 53, 122–124, 147
preference adjustment 69–70, 73
procreation: entitlements 71–72; duties to limit 93–97; financial costs of 123–124; relation to happiness 96–97
procreative liberty *see* reproductive freedom
pronatalism 70, 73, 78, 135
Purves, Duncan 43

Rachels, Stuart 122–124
Rawls, John 119–120
resource scarcity 57; *see also* population, and technological progress
reflective equilibrium 7–9
reproductive freedom 64, 72, 97–98, 131, 135–138
Rieder, Travis 87–88, 133
rights 117, 171; to bodily autonomy 133–134; human 63, 74, 77–78, 80n12, 136, 176, 176n9; to non-foreclosed opportunities 172; reproductive 97–98, 131–136

sentience 68
sex education 64–67
sex selection 74, 79n7, 94–95
Shiffrin, Seana 116–118
Simon, Julian 56–59
Singer, Peter 122–123
social discount rate 42–45

Time-Dependence Claim 39–40

United Nations International Conference on Population and Development 2–3, 6, 137
Universal Declaration of Human Rights 131
utilitarianism 7, 12n8, 124

virtue ethics 7–8, 174; and parenting 173–174

Warren, Mary Anne 68
Wasserman, David 127n10
Weinberg, Rivka 174–175
water shortage 15–16
Williams, Bernard 124

Printed in the United States
by Baker & Taylor Publisher Services